SpringerBriefs in World Mineral Deposits

Editor-in-chief

Antoni Camprubí, Instituto de Geología, Universidad Nacional Autónoma de México, México, Distrito Federal, México

Series editors

José María González Jiménez, Departamento de Geología, Universidad de Chile, Santiago de Chile, Chile
Francisco Javier González, Recursos Geológicos Marinos, Instituto Geológico y Minero de España, Madrid, Spain
Leo J. Millonig, Würzburg, Germany
John F. Slack, U.S. Geological Survey (Emeritus), Farmington, ME, USA

The *SpringerBriefs in World Mineral Deposits* book series seeks to publish monographs or case studies focused on a single mineral deposit or a limited group of deposits (sub-regional level), with regard to their mineralogy, structure, geochemistry, fluid geochemistry, and any other aspect that contributes to explaining their formation. This series is aimed at academic and company researchers, students, and other readers interested in the characteristics and creation of a certain deposit or mineralized area. The series presents peer-reviewed monographs.

The Springer Briefs in World Mineral Deposits series includes both single and multi-authored books. The Series Editors, Prof. Antoni Camprubí (UNAM, Mexico), Dr. Francisco Javier González, Dr. Leo Millonig, Dr. John Slack (SGA publications editor) and Dr. José María González Jiménez are currently accepting proposals and a proposal form can be obtained from our representative at Springer, Dr. Alexis Vizcaino (Alexis.Vizcaino@springer.com).

More information about this series at http://www.springer.com/series/15086

Rubén Piña

The Ni-Cu-(PGE) Aguablanca Ore Deposit (SW Spain)

 Springer

Rubén Piña
Department of Mineralogy and Petrology,
 Faculty of Geological Sciences
University Complutense of Madrid
Madrid
Spain

ISSN 2509-7857   ISSN 2509-7865   (electronic)
SpringerBriefs in World Mineral Deposits
ISBN 978-3-319-93153-1   ISBN 978-3-319-93154-8   (eBook)
https://doi.org/10.1007/978-3-319-93154-8

Library of Congress Control Number: 2018943253

Printed on acid-free paper

This Springer imprint is published by the registered company Springer International Publishing AG
part of Springer Nature
The registered company address is: Gewerbestrasse 11, 6330 Cham, Switzerland

To Dolores

Acknowledgements

Many people have contributed directly or indirectly to this book. I started working on Aguablanca almost 20 years ago. It was the subject of my Ph.D. Thesis. Rosario Lunar, Lorena Ortega, and Fernando Gervilla were my mentors during that early stage and I owe them my first thanks. Their trust and friendship have been my engine during all these years. Since then, and although I have been involved in other many research topics during all this time, Aguablanca has always been in my thoughts representing my main line of research. I will be direct. This book had not been possible without the collaboration of many colleagues from different Universities and Institutions. The list would be quite long, and if I put it whole, I would risk leaving someone out. However, I cannot stop mentioning Fernando Gervilla (who convinced me to write this book), Lorena Ortega, Rosario Lunar, Ramón Capote, Cecilio Quesada, Ignacio Romeo, Tuomo Alapieti (rest in peace), Sarah-Jane Barnes, Dany Savard, and Steve J. Barnes (who in addition kindly reviewed critically a first draft of the book). My research on Aguablanca progressed thanks to the input, support, and collaboration of all these colleagues. Importantly, I would like to thank the mine geologists, in particular César Martínez and Casimiro Maldonado, and their companies, for sharing your knowledge with me, providing samples, and supporting the research. Finally, thanks to Alexis Vizcaíno and José María Gonzalez-Jimenez who did possible this book with their edition.

Contents

List of Figures

Chapter 1
Introduction

1.1 Aguablanca, an Uncommon Example of Ni–Cu Sulfide Ore Deposit

The Aguablanca ore deposit (Lat. 37° 57′N, Long. 6° 11′W, Badajoz province, SW Spain) represents the first and unique example to date of an economic Ni–Cu–(PGE) magmatic sulfide deposit related to mafic-ultramafic magmatism in southern Europe. Its discovery by Presur-Atlantic Copper S.A. in 1993 resulted an unexpected finding because it was found in the European Iberian chain, a collisional orogenic belt in where no other Ni–Cu sulfide deposit has been documented to date. It is one of few deposits of this type located in Europe. With the exception of the world-class Noril'sk and Pechenga deposits in Russia, most Ni–Cu sulfide mineralizations are restricted to Finland that hosts the Kevitsa open-pit mine (237 Mt at 0.28% Ni, 0.41% Cu and 0.6 ppm PGM+Au, Santaguida et al. 2015), several projects under exploration (e.g., Sakatti, Coppard et al. 2013; Brownscombe et al. 2015) and numerous contact- and reef-type PGE occurrences (Makkonen et al. 2017). Other minor known examples in Europe include the Bruvann deposit in the Caledonide Råna intrusion in north Norway (Boyd and Mathiesen 1979) and Ivrea-Verbano in Italy (Garuti et al. 1986; Sessa et al. 2017).

Several features make Aguablanca a particularly interesting case of Ni–Cu sulfide deposit. Firstly, Aguablanca occurs in an uncommon geodynamic context for this type of ore mineralization, an orogenic belt developed in a convergent plate margin setting. Traditionally, these compressional settings have been commonly neglected in their potential for hosting Ni–Cu sulfide ores in favor of rifted intracontinental environments due mainly to theoretical considerations. Since 1980s, most world-class Ni sulfide deposits were observed to be related to intra-plate rifting (e.g., Noril'sk, Duluth, Pechenga, Kambalda, Naldrett 2004), and it was assumed that rifted environments with voluminous emplacement of mafic or ultramafic mantle-derived magmas were scenarios necessary for the formation of magmatic sulfide deposits. As a consequence, arc-related orogenic settings remained underexplored during many

© The Author(s) 2019
R. Piña, *The Ni-Cu-(PGE) Aguablanca Ore Deposit (SW Spain)*,
SpringerBriefs in World Mineral Deposits, https://doi.org/10.1007/978-3-319-93154-8_1

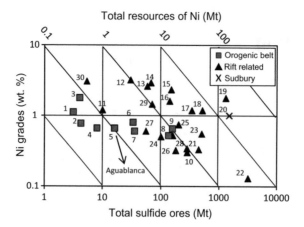

Fig. 1.1 Plot of Ni grade (wt%) versus size of deposit represented as total sulfide ores in million tonnes (Mt) for Ni–Cu–PGE deposits worldwide occurring in orogenic belt and rift-related settings. 1 Americano do Brazil, 2 Giant Mascot, 3 Sally Malay, 4 Vammala, 5 Aguablanca, 6 Kalatongke, 7 Heishan, 8 Huangshandong, 9 Xiarihamu, 10 Kevitsa, 11 Munali, 12 Raglan, 13 Kabanga, 14 Kambalda, 15 Thompson, 16 Voiseys' Bay, 17 Pechenga, 18 Jinchuan, 19 Noril'sk, 20 Sudbury, 21 Nebo-Babel, 22 Duluth, 23 Mount Keith, 24 Le Perouse, 25 Leinster Camp, 26 Abitibi, 27 Ntaka Hill 28 Uitkomst 29 Niquelandia, 30 Eagle

years. However, this situation has changed as result of the discovery of several arc setting–related Ni–Cu sulfide ores in the last 10–15 years (Nixon et al. 2015 and reference therein). Currently, examples of these deposits include the Sally Malay (now Savannah) intrusion in Australia (Hoatson and Blake 2000), the Tati and Selebi-Phikwe deposits in Botswana (Maier et al. 2008), the Giant Mascot deposit in British Columbia, Canada (Manor et al. 2016), the deposit of the Vammala and Kotalathi belts of Finland (Makkonen et al. 2008), the Xiarihamu deposit, the largest deposit found in an arc setting worldwide to date, in the East Kunlun orogenic belt in China (Li et al. 2015) and the Aguablanca deposit (Tornos et al. 2001; Piña et al. 2010). Although Ni–Cu sulfide deposits from orogenic belts are generally smaller in size and have lower Ni grades than those occurring in rift-related settings (Fig. 1.1), they highlight that the traditionally assumed scenario of Ni sulfide ores linked to intra-plate rifting environments is too restricted, and subduction-related orogenic belts must be considered as potentially favorable regions. The discovery of the world-class Xiarihamu deposit in 2011 (the second largest Ni deposit in China) has been an encouraging example for the exploration of Ni–Cu magmatic sulfides in arc settings resulting in the identification of sulfide mineralizations in other subduction-related mafic-ultramafic intrusions (e.g., Erbutu and Beiligaimiao, Peng et al. 2013, 2017).

Another outstanding feature of Aguablanca is that the sulfide mineralization is concentrated in a chaotic subvertical magmatic breccia, where variable proportions of sulfides and silicates form a matrix to unmineralized igneous and metasedimentary rock fragments probably derived from underlying igneous cumulates and country host rocks, respectively (Tornos et al. 2001; Piña et al. 2006). Although these breccia-textured ores have been recognized and described in other deposits (e.g., Voisey's Bay, Li and Naldrett 2000; Barnes et al. 2017; Nebo-Babel, Seat et al. 2007), they have been generally understudied and many features of their formation remain poorly understood, as for example, the emplacement mechanism. Most researchers have assumed that sulfide-matrix breccias are injected upward from their deep accumulation sites to their shallower levels of emplacement. However, it is not clear which mechanism/s can transport upward a dense sulfide liquid through much less dense crustal rocks. Deformation along major faults, seismic pumping, and structural readjustments are some of the invoked mechanisms. In Aguablanca, geophysical and structural studies suggest that the combination of opening of subvertical pull-apart conduits within a transpressional regimen and overpressured dense magma may have been the driving force of upward emplacement of slurries composed of sulfide liquid, rock fragments and silicate magma (Romeo et al. 2008). Recently, Barnes et al. (2018) have suggested an alternative model for this type of sulfide-matrix breccias consisting in gravity-driven percolation downward of molten sulfide through a pre-existing silicate-matrix intrusion breccia. According to this model, the sulfide melt would penetrate into the neck of the funnel-shaped Aguablanca intrusion through the intra-clast pore space of a pre-existing breccia formed by igneous and country-rock fragments in gabbronorite matrix, partially melting the silicate clasts and displacing upward the silicate melt component.

The particularities indicated above show that Aguablanca is an unusual ore deposit, not only in the metallogeny of Europe but also for Ni–Cu–(PGE) sulfide ore deposits. The objective of this monograph about Aguablanca is to compile the current state of knowledge of the deposit, highlighting important ore-forming processes.

1.2 Discovery and Mining History of Aguablanca

The Aguablanca mineralization was discovered in 1993 by Presur-Atlantic Copper S.A. (a joint venture between the Spanish State company Presur and Atlantic Copper S.A., formerly Rio Tinto Minera S.A.) during a regional exploration program that revealed Ni geochemical anomalies related to a gossan developed on altered gabbros (Lunar et al. 1997; Ortega et al. 2000). From this discovery, the company carried out detailed geological and geophysical work, including a program of short (less than 5 m) vertical percussion drillholes and long (up to 550 m) 45° dip diamond core holes, to trace the extension of the oxidized cap and to identify sulfide mineralization at depth. The shape and extension of orebodies were then delineated from a more exhaustive drilling program with more than 33.000 m of core. Estimated, proven and probable reserves of 15.7 Mt, grading 0.66 wt% Ni, 0.46 wt% Cu, 0.47 g/t PGM and 0.13 g/t Au were established. The deposit was then acquired by Rio Narcea Gold Mines Ltd in November of 2001, which operated the mine as open-pit with an on-site processing plant from 2004 until the end of 2007 when the mine was sold to Lundin

Mining. This company continued the mining activities with an annual production of ~8000 t Ni and 7000 t Cu (www.lundinmining.com) until the end of 2015 when the open-pit exhausted. A small underground mine with a life of 3.5 years was planned to operate from 2016 to recover 3.2 Mt of mineralized rock, but problems with the Environment Impact Declaration and a notable decrease in Ni and Cu prices triggered that the Aguablanca mine was indefinitely closed at the beginning of 2016. In November 2016, Valoriza Minera, a subsidiary of Sacyr, acquired Aguablanca, planning to start the mining production as underground mine as soon as possible.

1.3   Regional Exploration for Ni–Cu Sulfide Ores

The discovery of Aguablanca promoted an extensive and ambitious exploration program in the region, which led to the identification of more than 100 targets with radiometric anomalies similar to those of Aguablanca, one third of these targets having favorable mafic to ultramafic intrusive rocks, such as Argallón, Brovales and Cortegana (Martín-Izard et al. 2006; Piña et al. 2012). Some prospects show outcropping olivine-rich gabbros with visible magmatic sulfides and geochemical evidence of silica contamination that is considered as an important factor enhancing the potential for sulfide mineralization. This is the case of the Cortegana Igneous Complex, an assemblage of small mafic intrusive bodies located in the Aracena Metamorphic Belt, where some cores intercepted disseminated sulfide-bearing gabbronorites with up to 1.36 wt% Ni and 0.2 wt% Cu (Piña et al. 2012). Despite good initial results, exploration work showed that the sulfide mineralization at Cortegana is much less abundant than in Aguablanca and the project did not progress. This same situation can be extrapolated to other small mafic-ultramafic intrusive bodies throughout the Ossa-Morena Zone where either the sulfide mineralization is very scarce or the intrusion is not mineralized at all. Nevertheless, the Aguablanca experience tells us that the Ossa Morena Zone hosts numerous mafic to ultramafic stocks with potential for Ni–Cu sulfide mineralization. The dramatic economical and financial crisis from late 2000s has almost totally paralyzed exploration in the region, so it is certain that the potential of the region for hosting Ni–Cu ores has not been exhaustively evaluated.

References

Barnes SJ, Le Vaillant M, Lightfoot PC (2017) Textural development in sulfide-matrix ore breccias in the Voisey's Bay Ni–Cu–Co deposit, Labrador, Canada. Ore Geol Rev 90:414–438

Barnes SJ, Piña R, Le Vaillant M (2018) Textural development in sulfide-matrix ore breccias in the Aguablanca Ni–Cu deposit, Spain, revealed by X-ray fluorescence microscopy. Ore Geol Rev 95:849–862

Boyd R, Mathiesen CO (1979) The nickel mineralization of Råna mafic intrusion, Nordland, Norway. Can Mineral 17:287–298

Brownscombe W, Ihlenfeld C, Coppard J, Hartshorne C, Klatt S, Siikaluoma JK, Herrington RJ (2015) The Sakatti Cu–Ni–PGE sulfide deposit in Northern Finland. In: Maier WD, Lahtinen R, O'Brien H (eds) Mineral deposits of Finland. Elsevier, Oxford, pp 211–252

Coppard J, Klatt S, Ihlenfeld C (2013) The Sakatti Ni–Cu–PGE deposit in northern Finland. In: Hanski E, Maier W (eds) Excursion Guidebook FINRUS, Ni–Cr–PGE Deposits of Finland and the Kola Peninsula. 12th SGA Biennial Meeting, Mineral Deposit Research for a High-Tech World. Geological Survey of Sweden, Uppsala, pp 10–13

Garuti G, Fiandri P, Rossi A (1986) Sulphide composition and phase relations in the Fe–Ni–Cu ore deposits of the Ivrea-Verbano basic complex (western Alps, Italy). Mineral Deposita 21:22–34

Hoatson DM, Blake DH (2000) Geology and economic potential of the Paleoproterozoic layered mafic-ultramafic intrusions in the East Kimberley, Western Australia. In: Hoatson DM, Blake DH (eds) Australian Geological Survey Organization Bulletin, vol 246, p 496

Li C, Naldrett AJ (2000) Melting reactions of gneissic inclusions with enclosing magma at Voisey's Bay, Labrador, Canada; implications with respect to ore genesis. Econ Geol 95:801–814

Li C, Zhang Z, Li W, Wang Y, Sun T, Ripley EM (2015) Geochronology, petrology and Hf-S isotope geochemistry of the newly-discovered Xiarihamu magmatic Ni–Cu sulfide deposit in the Qinghai-Tibet plateau, western China. Lithos 216–217:224–240

Lunar R, García-Palomero F, Ortega L, Sierra J, Moreno T, Prichard H (1997) Ni–Cu–(PGM) mineralization associated with mafic and ultramafic rocks: The recently discovered Aguablanca ore deposit, SW Spain. In: Papunen H (ed) Mineral deposits: research and exploration: where do they meet?. Balkema, Rotterdam, pp 463–466

Maier WD, Barnes SJ, Chinyepi G, Barton JM, Eglington B, Setshedi I (2008) The composition of magmatic Ni–Cu–(PGE) sulfide deposits in the Tati and Selebi-Phikwe belts of eastern Botswana. Mineral Deposita 43:37–60

Makkonen HV, Mäkinen J, Kontoniemi O (2008) Geochemical discrimination between barren and mineralized intrusions in the Svecofennian (1.9 Ga) Kotalathi Nickel Belt, Finland. Ore Geol Rev 33:101–114

Makkonen HV, Halkoaho T, Konnunaho J, Rasilainen K, Kontinen A, Eilu P (2017) Ni–(Cu-PGE) deposits in Finland—geology and exploration potential. Ore Geol Rev 90:667–696

Manor MJ, Scoates JS, Nixon GT, Ames DE (2016) The Giant Mascot Ni–Cu–PGE deposit, British Columbia: mineralized conduits in a convergent margin tectonic setting. Econ Geol 111:57–87

Martín-Izard A, Fuertes M, Cepedal M, Rodríguez-Pevida L, Luceño L, Rodríguez D, Videira JC (2006) Reacciones de asimilación de rocas pelíticas en el proceso de formación de las mineralizaciones de Ni–Cu de Argallón, Cortegana y Olivenza (Ossa–Morena). Macla 6:297–298

Naldrett AJ (2004) Magmatic sulfide deposit: geology, geochemistry and exploration. Springer, Berlin, p 728

Nixon GT, Manor MJ, Jackson-Brown S, Scoates JS, Ames DE (2015) Magmatic Ni–Cu–PGE deposits at convergent margins. In: Ames DE, Houlé MG (eds) Targeted Geoscience Initiative 4: Canadian nickel-copper-platinum group element-chromium ore systems—fertility, pathfinders, new and revised models. Geological Survey of Canada, Open File 7856, pp 17–34

Ortega L, Prichard H, Lunar R, García-Palomero F, Moreno T, Fisher P (2000) The Aguablanca discovery. Min Mag 2:78–80

Peng R, Zhai Y, Li C, Ripley EM (2013) The Erbutu Ni–Cu deposit in the Central Asian Orogenic Belt: a Permian magmatic sulfide deposit related to boninitic magmatism in an arc setting. Econ Geol 108:1879–1888

Peng R, Li C, Zhai Y, Ripley EM (2017) Geochronology, petrology and geochemistry of the Beiligaimiao magmatic sulfide deposit in a Paleozoic active continental margin, North China. Ore Geol Rev 90:607–617

Piña R, Lunar R, Ortega L, Gervilla F, Alapieti T, Martínez C (2006) Petrology and geochemistry of mafic-ultramafic fragments from the Aguablanca (SW Spain) Ni–Cu ore breccia: Implications for the genesis of the deposit. Econ Geol 101:865–881

Piña R, Romeo I, Ortega L, Lunar R, Capote R, Gervilla F, Tejero R, Quesada C (2010) Origin and emplacement of the Aguablanca magmatic Ni–Cu–(PGE) sulfide deposit, SW Iberia: a multidisciplinary approach. Geol Soc Am Bull 122:915–925

Piña R, Gervilla F, Ortega L, Lunar R (2012) Geochemical constraints on the origin of the Ni–Cu sulfide ores in the Tejadillas prospect (Cortegana Igneous Complex), SW Spain. Res Geol 62:263–280

Romeo I, Tejero R, Capote R, Lunar R (2008) 3D gravity modeling of the Aguablanca Stock, tectonic control and emplacement of a Variscan gabbronorite bearing a Ni–Cu–PGE ore, SW Iberia. Geol Mag 145:345–359

Santaguida F, Luolavirta K, Lappalainen M, Ylinen J, Voipio T, Jones S (2015) The Kevitsa Ni–Cu–PGE deposit in the Central Lapland greenstone belt in Finland. In: Maier WD, Lahtinen R, O'Brien H (eds) Mineral deposits of Finland. Elsevier, Oxford, pp 195–210

Seat Z, Beresford SW, Grguric BA, Waugh RS, Hronsky JMA, Gee MAM, Groves DI, Mathison CI (2007) Architecture and emplacement of the Nebo-Babel gabbronorite-hosted magmatic Ni–Cu–PGE sulphide deposit, West Musgrave, Western Australia. Mineral Deposita 42:551–581

Sessa G, Moroni M, Tumiati S, Caruso S, Fiorentini ML (2017) Ni–Fe–Cu–PGE ore deposition driven by metasomatic fluids and melt-rock reactions in the deep crust: the ultramafic pipe of Valmaggia, Ivrea-Verbano, Italy. Ore Geol Rev 90:485–509

Tornos F, Casquet C, Galindo C, Velasco F, Canales A (2001) A new style of Ni–Cu mineralization related to magmatic breccia pipes in a transpressional magmatic arc, Aguablanca, Spain. Mineral Deposita 36:700–706

Chapter 2
Geological Setting

2.1 The Ossa-Morena Zone

The Aguablanca Ni–Cu sulfide ores occur within the Aguablanca stock, a small (approximately 3 km^2 in area) subcircular mafic intrusion located in the northern part of the Santa Olalla Igneous Complex (SOIC). The SOIC is located in the southern limb of the Olivenza-Monesterio antiform, a major WNW-ESE trending Variscan structure, occupying a central position within the Ossa-Morena Zone (OMZ) (Fig. 2.1). This zone comprises one of the SW divisions of the Iberian Massif, which corresponds to the westernmost outcrops of the Variscan orogen in Europe. A detailed geological review of the Iberian Massif is given by Quesada (1991) and references therein.

The OMZ is located between the Central Iberian Zone (CIZ), to the north, and the South Portuguese Zone (SPZ), to the south (Fig. 2.1a). The boundary between the OMZ and the CIZ is marked by the Badajoz-Córdoba shear zone, a tectonic unit interpreted as a Cadomian suture reactivated as an intraplate shear zone during the Variscan orogeny (Quesada 1991; Ábalos et al. 1991; Eguíluz et al. 2000) or a real Variscan suture (Azor et al. 1994; Simancas et al. 2001). The contact with the SPZ, to the south, is marked by the Beja-Acebuches amphibolites and the Pulo do Lobo accretionary prism. These units have been interpreted as remnants of the Rheic Ocean accreted to the southern margin of the OMZ during the oblique Variscan convergence. These remnants delineate the Variscan suture between Gondwana and Laurussia during the formation of Pangaea (Quesada et al. 1994; Castro et al. 1996; Braid et al. 2010).

The structural evolution of the OMZ during the Variscan orogeny (from Middle Devonian to Early Permian) was mainly governed by transpressional tectonics. This transpressional regime resulted in the formation of a dominant thick-skinned transcurrent architecture delineated by oblique thrust nappes, folds, and strike-slip fault associations (Expósito 2000). Internal deformation of each horst is variable and includes several folding and oblique thrust generations. The Olivenza-Monesterio

© The Author(s) 2019
R. Piña, *The Ni-Cu-(PGE) Aguablanca Ore Deposit (SW Spain)*,
SpringerBriefs in World Mineral Deposits, https://doi.org/10.1007/978-3-319-93154-8_2

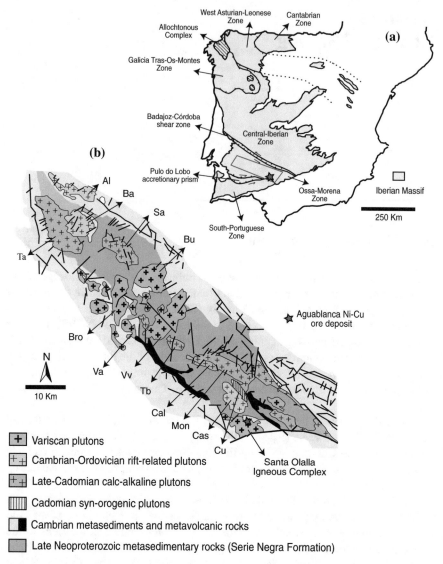

Fig. 2.1 a Sketch of the Iberian Peninsula showing the main zones of the Iberian Massif and the location of the Olivenza-Monesterio antiform that hosts the Aguablanca ore deposit; **b** Simplified geological map of the Olivenza-Monesterio antiform showing outcrop areas of the main igneous bodies and the location of the Aguablanca Ni–Cu ore deposit in the north boundary of the Santa Olalla Igneous Complex. Cu: Culebrín, Cas: Castillo, Mon: Monesterio, Cal: Calera de León, Tb: Tablada, Vv: Valencia del Ventoso, Va: Valuengo, Bro: Brovales, Ta: Táliga, Al: Almendral, Ba: Barcarrota, Sa: Salvatierra, Bu: Burguillos del Cerro

antiform is a major Variscan structure in the OMZ. It was generated by basement-involved ductile thrusting which gave rise to a basement antiformal stack, while the

Paleozoic cover detached from the basement and initially formed a typically thin-skinned, SW-verging imbricate thrust fan and regional associated recumbent folds to the south (Vauchez 1975; Quesada et al. 1994; Expósito 2000). A second folding event, also SW-vergent but characterized by steep axial planes, affected the already deformed thin-skinned imbricate thrust fan where some syn-orogenic basins were formed (e.g., the Terena flysch basin). Transpression is indicated by existence of a major sinistral component during the entire evolution in the main NW-SE thrust structures, anticlockwise fold transection and associated (NE-SW trending) synthetic and (NNW-SSE trending) antithetic Riedel faults. The sigmoidal map pattern that characterizes the OMZ represents the cumulative result of all these elements.

2.2 Magmatism

The complex tectonic evolution of the OMZ was accompanied by intense magmatism which can be divided into three main stages well represented in the Olivenza-Monesterio antiform (Fig. 2.1b): (1) subduction-related magmatism during the Neoproterozoic Cadomian orogeny, (2) Cambrian-Ordovician magmatism related to rifting leading to the opening of oceanic crust (Rheic ocean), and (3) Carboniferous Variscan collision- and extension-related magmatism. The main magmatism related to the Cadomian orogeny corresponds to calc-alkaline volcanic and plutonic rocks (mainly, amphibolites and diorite-granites), showing a typical subduction arc signature and dated in the range c. 587–532 Ma (Schäfer 1990; Ochsner 1993; Quesada 1990). The Cambrian-Ordovician rifting event was accompanied by bimodal igneous activity of tholeiitic and alkaline affinity (Mata and Munhá 1990; Sánchez-García et al. 2003, 2010) in the range c. 530–470 Ma (Ochsner 1993; Ordóñez-Casado 1998; Montero et al. 2000). The last main magmatic event recorded in the OMZ took place during the Variscan orogeny. It was characterized by intermediate to felsic calc-alkaline compositions ranging from metaluminous tonalite and granodiorite to peraluminous granite and leucogranite, with volumetrically minor gabbros. Numerous sub circular plutons were emplaced throughout the Olivenza-Monesterio antiform in the range c. 350–330 Ma (Dallmeyer et al. 1995; Montero et al. 2000; Pin et al. 2008) (Fig. 2.1b): Valencia del Ventoso, Bazana, Brovales, Valuengo and Burguillos del Cerro. Separated from this group of plutons, 50 km to the SE, is the Santa Olalla Igneous Complex that includes the Aguablanca stock, the small mafic igneous body that hosts the Aguablanca Ni–Cu sulfide ores.

2.3 The Iberian Reflective Body (IRB)

A deep seismic reflection profile (IBERSEIS) running 303 km long across the northern South Portuguese Zone, Ossa-Morena Zone and southern Central Iberian Zone revealed the existence of a 140 km-long and up to 5 km-thick high amplitude

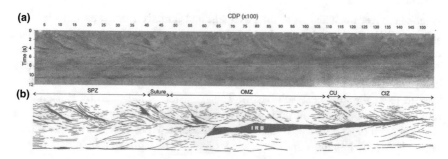

Fig. 2.2 **a** Time migrated stack image of the IBERSEIS deep seismic reflection profile. **b** Line drawing interpretation of the seismic reflection profile showing the location of the main tectonic units: SPZ, South Portuguese Zone; OMZ, Ossa-Morena Zone; CIZ, Central Iberian Zone; Suture, accretionary complex between SPZ and OMZ; CU, accretionary wedge between OMZ and CIZ. The green body indicates the location and geometry of the Iberian Reflective Body (IRB) (Reprinted from Simancas et al. 2006, with permission from The Geological Society of London)

reflectivity body, called the Iberian Reflective Body, at about 10–15 km depth throughout the Ossa-Morena Zone (Fig. 2.2; Simancas et al. 2003). From its relationships with major structures, the Iberian Reflective Body has been interpreted to represent a large layered mafic-ultramafic intrusion formed during the activity of a syn-orogenic mantle plume in Early Carboniferous times (~340–350 Ma) (Simancas et al. 2003; Palomeras et al. 2011; Brown et al. 2012). According to Simancas et al. (2003, 2006) and Carbonell et al. (2004), this mantle plume was likely active during a short-lived intra-orogenic extensional event intermediate between two main transpressional tectonic regimes associated with a subduction-related magmatic arc (Fig. 2.3; Simancas et al. 2009). Alternatively, Tornos and Casquet (2005) invoked for the Iberian Reflective Body an origin related to regional compression decollement, which became a zone of major crustal decoupling resulting in the injection of juvenile magmas. Almost simultaneously to the discovery of the Iberian Reflective Body, Pous et al. (2004) found a high-conductivity layer at a depth of 15–20 km below the whole Ossa-Morena Zone spatially correlated with the Iberian Reflective Body. These authors concluded that the mafic-ultramafic intrusions could not by themselves explain the high conductivity of the layer, because igneous rocks typically exhibit low conductivity. Alternatively, Pous et al. (2004) suggested the presence of conductive graphite-rich shale screens, likely belonging to the Precambrian Serie Negra Formation, intercalated among multiple sheet-like mafic intrusions. More recently, the existence of this mafic-ultramafic magmatism in the middle crust of the Ossa Morena Zone has been also supported by the geophysical studies carried out by Palomeras et al. (2011) and Brown et al. (2012), such that the emplacement of voluminous mafic-ultramafic magmas in the middle crust of the Ossa-Morena Zone in Early Carboniferous times is well established. Tornos et al. (2006) have suggested that dismembered parts of this mafic-ultramafic rocks crop out in the Aracena Massif (e.g., Cortegana Igneous Complex) located adjacent to the south Portuguese Zone.

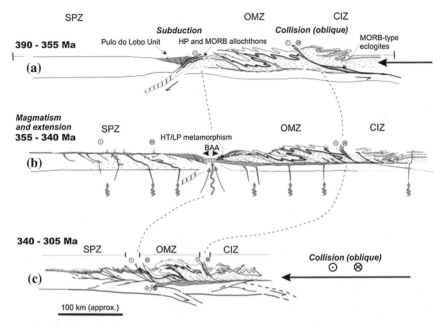

Fig. 2.3 Schematic cartoon of the geological evolution of the SW Iberia during the Variscan orogeny (Middle Devonian—Late Carboniferous) showing the emplacement of the Iberian Reflective Body (IRB, green) during an intra-orogenic extensional event intermediate between two main collisional events (Reproduced from Simancas et al. 2009. Copyright © 2008 Académie des sciences, published by Elsevier Masson SAS. All rights reserved)

Several studies have highlighted the regional implications of the emplacement of this huge volume of magmas in the Ossa Morena Zone. Simancas et al. (2003) have suggested that gabbros and diorites intruded in the OMZ during the Variscan orogeny (at ca. 350 Ma) probably came from the Iberian Reflective Body magmatic chamber. Similarly, Cambeses et al. (2015) suggested that the Iberian Reflective Body magmatism promoted melting, assimilation and mixing processes between mafic and felsic magmas that ultimately resulted in the formation of hybrid magmas later emplaced into the OMZ upper crust (e.g., Valencia del Ventoso, Burguillos del Cerro and Santa Olalla de Cala). These authors have also pointed to the emplacement of the Iberian Reflective Body as the main heat source for the high-temperature, low-pressure metamorphism developed in the Évora-Aracena-Lora del Río metamorphic belt of the OMZ. Other studies genetically link the Iberian Reflective Body magmatism with the metallogeny in the OMZ (Simancas et al. 2003; Tornos and Casquet 2005; Tornos et al. 2006; Ordoñez-Casado et al. 2008). For example, Tornos and Casquet (2005) suggest that the source of magmas and fluids responsible of the Ni–Cu sulfide mineralization in Aguablanca and several occurrences of IOCG mineralization, respectively, in the Ossa Morena Zone, was probably the Iberian Reflective Body magmatism. Ordóñez-Casado et al. (2008) also suggest that the

Iberian Reflective Body below the south part of the Central Iberian Zone could be involved in the genesis of Ni–Cu sulfide mineralization in gabbros and norites located in the northern part of Extremadura region, far away from the Aguablanca deposit.

References

Ábalos B, Gil Ibarguchi JI, Eguiluz L (1991) Cadomian subduction/collision and Variscan transpression the Badajoz-Córdoba Shear Belt (SW Spain). Tectonophysics 199:51–72

Azor A, Lodeiro FG, Simancas JF (1994) Tectonic evolution of the boundary between the Central Iberian and Ossa-Morena zones (Variscan belt, southwest Spain). Tectonics 13:45–61

Braid JA, Murphy JB, Quesada C (2010) Structural analysis of an accretionary prism in a continental collisional setting, the Late Paleozoic Pulo do Lobo Zone, Southern Iberia. Gondwana Res 17:422–439

Brown D, Zhang X, Palomeras I, Simancas F, Carbonell R, Juhlin C, Salisbury M (2012) Petrophysical analysis of a mid-crustal reflector in the IBERSEIS profile, SW Spain. Tectonophysics 550–553:35–46

Cambeses A, Scarrow JH, Montero P, Molina JF, Moreno JA (2015) SHRIMP U–Pb zircon dating of the Valencia del Ventoso plutonic complex, Ossa-Morena Zone, SW Iberia: Early Carboniferous intra-orogenic extension-related 'calc-alkaline' magmatism. Gondwana Res 28:735–756

Carbonell R, Simancas F, Juhlin C, Pous J, Pérez-Estaún A, González-Lodeiro F, Muñoz G, Heise W, Ayarza P (2004) Geophysical evidence of a mantle derived intrusion in SW Iberia. Geophys Res Lett 31:L11601

Castro A, Fernández C, De La Rosa J, Moreno-Ventas I, Rogers G (1996) Significance of MORB-derived amphibolites from the Aracena Metamorphic Belt, Southwest Spain. J Petrol 37:235–260

Dallmeyer RD, García-Casquero JL, Quesada C (1995) Ar/Ar mineral age constraints on the emplacement of the Burgillos del Cerro Igneous Complex (Ossa–Morena Zone, SW Iberia). Boletín Geológico y Minero 106:203–214

Eguíluz L, Gil Ibarguchi JI, Ábalos B, Apraiz A (2000) Superposed Hercynian and Cadomian orogenic cycles in the Ossa Morena Zone and related areas of the Iberian Massif. Geol Soc Am Bull 112:1398–1413

Expósito I (2000) Evolución estructural de la mitad septentrional de la Zona de Ossa-Morena, y su relación con el límite Zona de Ossa-Morena/Zona Centroibérica. PhD thesis, Universidad de Granada, p 296

Mata J, Munhá J (1990) Magmatogénese de Metavulcanitos Cámbricos do Nordeste Alentejano: os stádios iniciais de rifting continental. Comunicacoes dos Servicos Geologicos de Portugal 76:61–89

Montero P, Salman K, Bea F (2000) New data on the geochronology of the Ossa-Morena Zone, Iberian Massif. Basement Tectonics 15:136–138

Ochsner A (1993) U–Pb geochronology of the Upper Proterozoic–Lower Paleozoic geodynamic evolution in the Ossa–Morena Zone (SW Iberia): constraints on timing of the Cadomian orogeny. PhD thesis, Swiss Federal Institute of Technology, Zurich, p 249

Ordóñez-Casado B (1998) Gechronological studies of the Pre-Mesozoic basement of the Iberian Massif: the Ossa–Morena zone and the allochthonous complexes within the Central Iberian zone. PhD thesis, Swiss Federal Institute of Technology, Zurich, p 235

Ordóñez-Casado B, Martín-Izard A, García-Nieto J (2008) SHRIMP-zircon U-Pb dating of the Ni–Cu–PGE mineralized Aguablanca gabbro and Santa Olalla granodiorite: Confirmation of an Early Carboniferous metallogenic epoch in the Variscan Massif of the Iberian Peninsula. Ore Geol Rev 34:343–353

Palomeras I, Carbonell R, Ayarza P, Fernández M, Simancas JF, Martínez Poyatos D, González Lodeiro F, Pérez Estaún A (2011) Geophysical model of the lithosphere across the Variscan Belt of SW Iberia: multidisciplinary assessment. Tectonophysics 508:42–51

Pin C, Fonseca PE, Paquette JL, Castro P, Matte P (2008) The ca. 350 Ma Beja Igneous Complex: a record of transcurrent slab break-off in the Southern Iberia Variscan belt? Tectonophysics 461:356–377

Pous J, Muñoz G, Wiebke H, Melgarejo JC, Quesada C (2004) Electromagnetic imaging of Variscan crustal structures in SW Iberia: the role of interconnected graphite. Earth Planet Sci Lett 217:435–450

Quesada C (1990) Precambrian successions in SW Iberia: their relationship to Cadomian orogenic events. In: D'Lemos RS, Strachan RA, Topley CG (eds) The Cadomian Orogeny. Geological Society of Special Publication London, pp 553–562

Quesada C (1991) Geological constraints on the Paleozoic tectonic evolution of tectonostratigraphic terranes in Iberian Massif. Tectonophysics 185:225–245

Quesada C, Fonseca P, Munhá J, Oliveira JT, Ribeiro A (1994) The Beja-Acebuches ophiolite (Southern Iberia Variscan fold belt): geological characterization and geodynamic significance. Boletín Geológico y Minero 105:3–49

Sánchez-García T, Bellido F, Quesada C (2003) Geodynamic setting and geochemical signatures of Cambrian-Ordovician rift-related igneous rocks (Ossa-Morena Zone, SW Iberia). Tectonophysics 365:233–255

Sánchez-García T, Bellido F, Pereira MF, Chichorro M, Quesada C, Pin C, Silva JB (2010) Rift-related volcanism predating the birth of the Rheic Ocean. Gondwana Res 17:392–407

Schäfer HJ (1990) Geochronological investigations in the Ossa–Morena Zone, SW Spain. PhD thesis, Swiss Federal Institute of Technology, Zurich, p 153

Simancas JF, Martínez Poyatos D, Expósito I, Azor A, González Lodeiro F (2001) The structure of a major suture zone in the SW Iberian Massif: the Ossa-Morena/Central Iberian contact. Tectonophysics 332:295–308

Simancas JF, Carbonell R, González Lodeiro F, Pérez Estaún A, Juhlin C, Ayarza P, Kashubin A, Azor A, Martínez Poyatos D, Almodóvar GR, Pascual E, Sáez R, Expósito I (2003) Crustal structure of the transpressional Variscan orogen of SW Iberia: SW Iberia deep seismic reflection profile (IBERSEIS). Tectonics 22(1–5):1–19

Simancas F, Carbonell R, González Lodeiro F, Pérez Estaún A, Juhlin C, Ayarza P, Kashubin A, Azor A, Martinez Poyatos D, Saez R, Almodovar GR, Pascual E, Flecha I, Marti D (2006) Transpresional collision tectonics and mantle plume dynamics: the Variscides of southwestern Iberia. In: Gee D, Stephenson RA (eds) European lithosphere dynamics. Geological Society of London, Memoirs, vol 32, pp 345–354

Simancas F, Azor A, Martínez-Poyatos D, Tahiri A, El Hadi H, González-Lodeiro F, Pérez-Estaún A, Carbonell R (2009) Tectonic relationships of Southwest Iberia with the allochthons of Northwest Iberia and the Moroccan Variscides. CR Geosci 341:103–113

Tornos F, Galindo C, Casquet C, Rodríguez Pevida L, Martínez C, Martínez E, Velasco F, Iriondo A (2006) The Aguablanca Ni–(Cu) sulfide deposit, SW Spain: geologic and geochemical controls and the relationship with a midcrustal layered mafic complex. Mineral Deposita 41:737–769

Tornos F, Casquet C (2005) A new scenario for related IOCG and Ni–(Cu) mineralization: the relationship with giant mid-crustal mafic sills, Variscan Iberian Massif. Terranova 17:236–241

Vauchez A (1975) Tectoniques tangentielles superposées dans le segment hercynien sud-ibérique: les nappes et plis couchés de la region d'Alconchel-Fregenal de la Sierra (Badajoz). Boletín Geológico y Minero 86:573–580

Chapter 3
The Santa Olalla Igneous Complex (SOIC)

3.1 Igneous Rocks

The SOIC is a subrounded (up to 7 km long) calc-alkaline plutonic group structurally located in a wedge bounded by two main faults (Romeo et al. 2006a, 2008): the Cherneca Fault to the north, and the Zufre Fault to the south (Fig. 3.1). The SOIC is formed by two main plutons, the Santa Olalla intrusion and the Aguablanca mafic stock. The Santa Olalla intrusion, the largest pluton of the complex, shows a reverse compositional zoning with amphibole-biotite quartzdiorite in its north side grading to tonalite in the centre and monzogranite toward the southern limit (Velasco 1976; Casquet 1980). Tonalite is the most abundant igneous rock type and consists of an equigranular, medium to coarse-grained leucocratic rock (Fig. 3.2a) composed of zoned plagioclase, quartz, biotite and amphibole locally containing relicts of pyroxene. The intrusion contains variably digested gabbro and leucogranite enclaves (Fig. 3.2a, b), synplutonic dykes and a wide variety of disequilibrium and reaction microstructures (Bateman et al. 1992).

The Aguablanca mafic stock, the host of the Ni–Cu sulfide ores, occurs in the northern part of the SOIC, adjacent to the Cherneca ductile shear zone (Fig. 3.3). The Aguablanca stock consists of a small (~3 km^2) subcircular intrusive body of dark medium-grained gabbroic rocks (Fig. 3.2c). In detail, the Aguablanca rocks comprise cumulate-textured hornblende-bearing gabbronorite and minor gabbro and norite, grading to the south into diorite and quartzdiorite. These rocks commonly host leucocratic enclaves (Fig. 3.2d) and, at the southern parts of the stock close to the contact with the Santa Olalla tonalite, gabbro shows mingling textures with felsic hybrid rocks (Fig. 3.2e). Most rocks of the Aguablanca intrusion contain no Fe–Ni–Cu sulfides, the sulfide mineralization being restricted to the magmatic breccia, although small sulfide patches can be locally found where the igneous rocks envelope partially-digested xenoliths of country sedimentary rocks with no visible sulfides (Fig. 3.2f).

© The Author(s) 2019   15
R. Piña, *The Ni-Cu-(PGE) Aguablanca Ore Deposit (SW Spain)*,
SpringerBriefs in World Mineral Deposits, https://doi.org/10.1007/978-3-319-93154-8_3

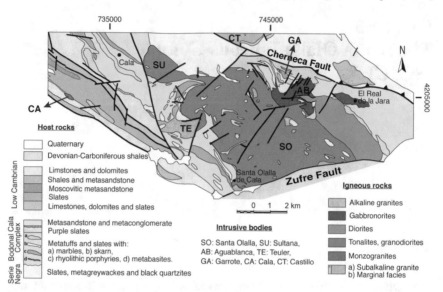

Fig. 3.1 Geological map of the Santa Olalla Igneous Complex showing the location of intrusive bodies and the main fault structures, Zufre and Cherneca. Abbreviations: AB, Aguablanca, SO, Santa Olalla; SU, Sultana; TE, Teuler; GA, Garrote; CA, Cala; CT, Castillo (Modified from printed figure by Romeo et al. (2006a), with permission from Elsevier)

The petrography and mineral chemistry of the Aguablanca igneous rocks have been described in detail by Piña et al. (2006). Hornblende-bearing gabbronorite is a medium to coarse-grained meso- and orthocumulate containing variable amounts of orthopyroxene (27–48 modal %, Mg# 0.85–0.78), plagioclase (23–47 modal %, An_{79-51}) and clinopyroxene (4–11 modal %, Mg# 0.89–0.83), and interstitial green-brown hornblende (10–21 modal %, Mg# 0.86–0.70), phlogopite (<5 modal %, Mg# 0.79–0.73) and minor quartz (<1 modal %) (Fig. 3.4a–d). Primary green-brown hornblende and phlogopite often include poikilitically pyroxene and plagioclase crystals (Fig. 3.4d). Locally, gabbronorites show subhorizontal layering at decimetre scale consisting of variations in the plagioclase modal contents. At the south, quartz diorite occurs as variably thick masses (up to 150 m) between gabbronorite with transitional contacts. The quartz diorite is composed of cumulus plagioclase (56–63 modal %, An_{45-37}), clinopyroxene (<5 modal %, Mg# 0.77–0.73) and orthopyroxene (<7 modal %, Mg# 0.67–0.62), and intercumulus phlogopite (7–18 modal %, Mg# 0.63–0.60), quartz (5–15 modal %) and amphibole (<7 modal %) (Fig. 3.4e, f). In all rock-types, primary mineralogical assemblages are variably altered to a secondary assemblage comprising actinolite, chlorite, bastite, talc, carbonates, serpentine, albite, sericite, and epidote-zoisite group minerals (Fig. 3.4g, h).

Around the SOIC, other minor granitic bodies occur: the Sultana, Teuler and Garrote intrusions (Fig. 3.1). Sultana, located at the NW of SOIC, consists of a small mafic apophysis composed of hornblende-biotite tonalite and quartzdiorite

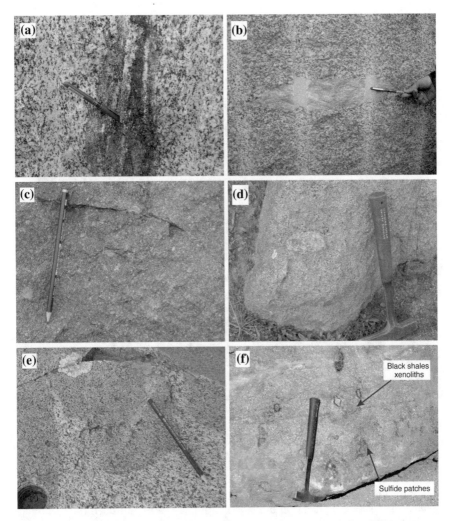

Fig. 3.2 Tonalite rocks from the Santa Olalla intrusion enveloping mafic (**a**) and leucratic (**b**) enclaves. **c** Dark gabbroic rocks from the Aguablanca stock. **d** Leucocratic enclave in Aguablanca mafic rocks. **e** Mingling texture between leucocratic and mafic igneous rocks in the bed of the Ribera de Cala River, close to the contact between Aguablanca and Santa Olalla intrusions. **f** Sulfide patches hosted by Aguablanca gabbroic rocks associated with partially digested xenoliths of black shales

(Apalategui et al. 1990). The Garrote intrusion is a small (<1 km^2) alkaline hornblende-bearing syenitic granite located close to the northern boundary of the Aguablanca stock. Finally, the Teuler intrusion is a fine-grained biotite-rich monzogranite located at the W of SOIC that has generated a magnesian skarn with stratiform magnetite mineralization (Tornos et al. 2004a). The SOIC includes some minor intrusion-related Cu–Au veins and a large calcic skarn adjacent to a minor apophysis to the SW (e.g., Cala; Velasco 1976; Velasco and Amigó 1981).

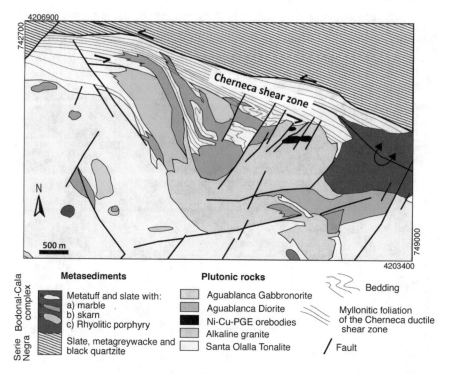

3.2   Host Rocks

The SOIC intruded in two different stratigraphic units regionally metamorphosed to greenschist facies (Fig. 3.1). In the NW margin, the SOIC intruded metasedimentary rocks composed of alternating pyrite-rich black slates and metagreywackes with minor intercalations of metavolcanic rocks and black quartzites (Fig. 3.5a). These rocks correspond to the Tentudía sucession, the upper part of the Late Neoproterozoic Serie Negra Formation, probably the most representative formation of the OMZ, reaching up to 15–25 km depth, which crops out along the central area of the Olivenza-Monesterio antiform. The Serie Negra Formation consists of different graphite- and pyrite-bearing rock-types including metacherts, quartz phyllilites, mica schist, metagreywackes and paragneises (Eguiluz 1988, Pous et al. 2004; Pereira et al. 2006). Schäfer et al. (1993) established a maximum age of deposition for this formation at 564 ± 9 Ma. Sulfur content in these rocks is variable but can reach up to 3000 ppm (Piña 2006). The S-rich black slates of the Serie Negra Formation are considered as the favourite contaminant providing the S necessary to the Aguablanca magma to

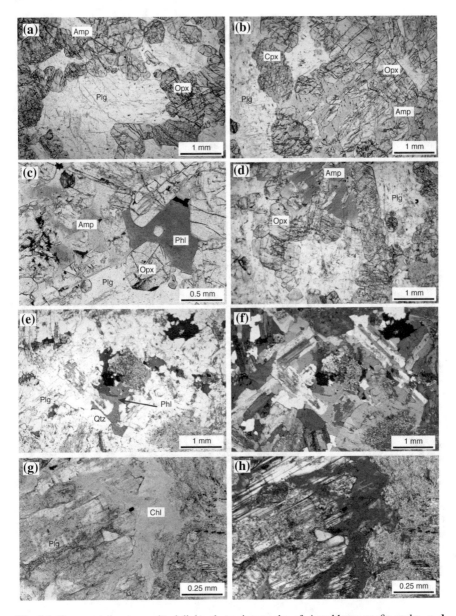

Fig. 3.4 Representative transmitted light photomicrographs of Aguablanca mafic rocks. **a–d** Hornblende-bearing gabbronorite showing euhedral to subhedral pyroxene and plagioclase crystals with interstitial amphibole and phlogopite grains. **e, f** Quartzdiorite with plagioclase, quartz and interstitial phlogopite (plane-polarized and cross-polarized light, respectively). **g, h** Partially-sericitized plagioclase replaced by chlorite along grain boundary in gabbronorite rock (plane-polarized and cross-polarized light, respectively)

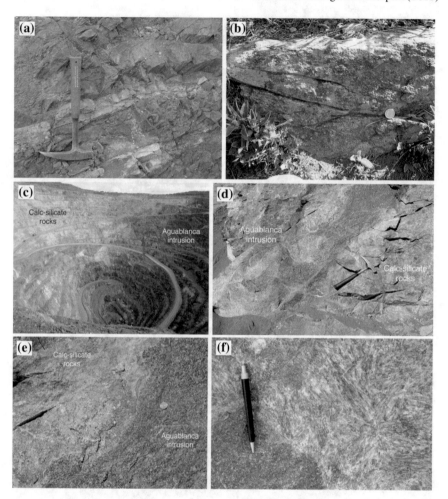

Fig. 3.5 **a** Intercalations of black slates and metagreywackes from the Tentudía sucession of the Late Neoproterozoic Serie Negra Formation. **b** Volcano-sedimentary rocks from the Cambrian Bodonal Cala Complex. **c** Open-pit of Aguablanca showing the contact between the calc-silicate rocks to the north with the ore-bearing Aguablanca intrusion to the south. **d, e** Detail of the intrusive contact between the calc-silicate rocks and the Aguablanca intrusion. **f** Skarn developed in the contact with the Aguablanca rocks composed of variable amounts of actinolite, epidote, garnet and scapolite

reach the sulfur saturation and, hence, form the sulfide mineralization (Casquet et al. 2001; Tornos et al. 2001; Piña et al. 2006).

At the E and W, the SOIC intruded a volcano-sedimentary sequence formed by tuffs and porphyritic rhyolites (Fig. 3.5b) of calc-alkaline affinity and variable-thick carbonate rocks at the top. This succession corresponds to the Cambrian Bodonal

Cala Complex (530 ± 3 Ma, U–Pb ID-TIMS on zircons, Romeo et al. 2006b) that uncomformably overlies the Serie Negra Formation.

The contact between the Aguablanca stock and Bodonal Cala rocks is well exposed in the open-pit (Fig. 3.5c). It is a well-defined, sharp intrusive contact (Fig. 3.5d, e). Along the northern contact with the Bodonal Cala carbonate rocks, the intrusion produced a well-developed exoskarn (~2 km wide) characterized by garnetite, marble and calc-silicate rocks (Fig. 3.5f) (Velasco 1976; Casquet 1980). The skarn mineralogy comprises scapolite, epidote, garnet, actinolite and clinopyroxene, with minor amounts of pyrite, pyrrhotite, sphalerite, chalcopyrite and galena. The marbles were affected by penetrative ductile deformation related to the sinistral transpressional kinematics of the Cherneca shear zone (Romeo et al. 2007). Decimetric-sized xenoliths of skarn, marble and calc-silicate rocks are locally found in the intrusive rocks near to the contact. Casquet (1980) estimated temperatures up to 750 °C (i.e., hypersthene hornfels facies) within the thermal aureole, near to the contact with the Aguablanca stock and inferred a depth of emplacement of 1.7–3.5 km from metamorphic mineral equilibria. The Santa Olalla intrusion shows numerous roof pendants of host country rocks scattered within igneous rocks, mostly skarnified limestone, interpreted as having come from the subhorizontal upper contact (Eguiluz et al. 1989).

3.3 Geochemistry

The SOIC is characterized by a high-K calc-alkaline affinity (Casquet et al. 2001; Piña et al. 2006), similar to other nearby intrusions such as Burguillos and Brovales. Rocks from the Santa Olalla intrusion and Aguablanca stock exhibit a well-defined trend of increasing in SiO_2, Al_2O_3, alkali and trace incompatible elements such as Rb and Ba with decreasing MgO (Fig. 3.6). FeO_t and Cr_2O_3 significantly decrease with decreasing MgO. Igneous facies of the Santa Olalla intrusion (namely, tonalite, diorite and monzogranite) have high SiO_2 (56–68 wt%), Al_2O_3, K_2O (>4.6 wt%), Ba, Sr and REE contents, and low MgO and Cr (<280 ppm) abundances and are hence more differentiated than the gabbronorites. In the case of the Aguablanca mafic stock, gabbronorite and norite rocks have more primitive compositions than diorites located in the southern part of the stock. In terms of major elements, there is no difference between mineralized and barren gabbronorites. Both gabbronorites contain relatively low SiO_2 contents (48–57 wt%), K_2O (<0.7 wt%), Ba, Rb, Th and Nb, and high MgO (12.2–16.2 wt%) and Cr (738–1200 ppm) abundances.

Trace element whole rock geochemistry of the sulfide-bearing gabbronorite and sulfide-free rocks (including gabbronorite and diorite) from the Aguablanca stock is quite similar (Fig. 3.7, Piña et al. 2006). In all rock-types, primitive mantle-normalized trace incompatible element patterns are characterized by enrichment in large-ion lithophile elements (LILE), Rb, Ba, Th and U, relative to high field strength elements (HFSEs), Nb, Ta, Zr and Hf, strong negative Nb and Ta and positive Eu and Sr anomalies (probably, reflecting plagioclase accumulation). Rocks are enriched in

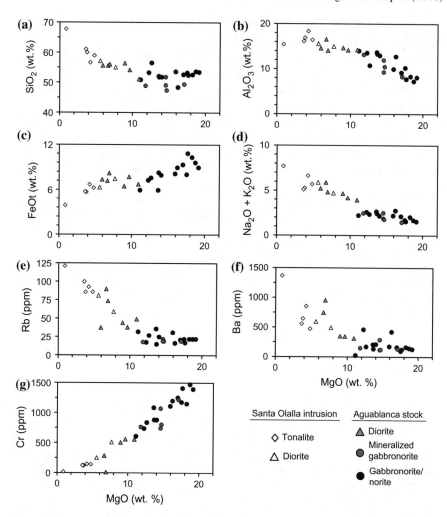

Fig. 3.6 Binary variation diagrams of major oxides (SiO_2, Al_2O_3, FeO_t, and $Na_2O + K_2O$, (**a–d**) and minor elements (Rb, Ba and Cr, **e–g**) versus MgO for rocks from the Santa Olalla and Aguablanca intrusions

LREE relative to HREE with $(La/Lu)_N$ ranging from 3.8 to 5.4 in gabbronorite. The abundance of trace incompatible elements is typically higher in rocks from the Santa Olalla intrusion than in those from Aguablanca stock (Fig. 3.7).

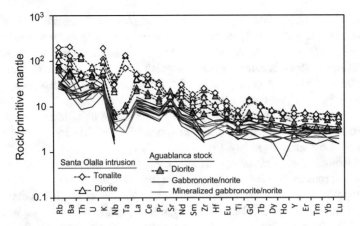

Fig. 3.7 Whole-rock primitive mantle-normalized trace element patterns of rocks from the Santa Olalla and Aguablanca intrusions

3.4 Timing Relations and Geochronology

Based on geochemical and petrological similarities with other nearby Variscan igneous complexes located in the Olivenza-Monesterio antiform such as Burguillos del Cerro, Brovales and Valencia del Ventoso, the SOIC is interpreted to be a product of the Variscan calc-alkaline magmatism developed in the Ossa-Morena Zone (Casquet 1980; Eguiluz et al. 1989; and Apalategui et al. 1990). The timing and genetic relationships between the Aguablanca stock and the Santa Olalla intrusion have been the subject of a number of studies. Casquet (1980) considered that Aguablanca gabbro was crosscut by the Santa Olalla tonalite, suggesting that Aguablanca predated the Santa Olalla intrusion. By contrast, Eguiluz et al. (1989) and Bateman et al. (1992) considered both intrusions as different igneous facies of a same post-tectonic intrusive body emplaced during Early Carboniferous. However, these works proposed different processes to explain the compositional variability. Eguiluz et al. (1989) proposed that the SOIC is an I-type igneous body whose reverse compositional zoning records a typical calc-alkaline evolution by fractional crystallization. By contrast, Bateman et al. (1992) suggested that the different igneous facies are not the result of single fractional crystallization and alternatively proposed that the tonalite facies are the result of mixing between S-type cordierite-rich magmas with basaltic magmas that previously fractionated giving rise to the Aguablanca mafic rocks. Several later studies based on the isotope Sr–Nd compositions and whole rock geochemistry of the SOIC proposed an evolution of the parental magmas by assimilation-fractional crystallization (AFC) processes (Casquet et al. 1998, 2001). According to these authors, the initial fractionation of pyroxene and olivine from parental basaltic magmas and the early assimilation of S-bearing metasediments resulted in the formation of early cumulate rocks and sulfide melts. Subsequently, the fractionated magma enriched in crustal components evolved by fractional crystallization of plagioclase and amphibole giving rise to the different igneous facies of Santa Olalla intrusion (i.e., diorite, tonalite and monzogranite).

From the discovery of the Ni–Cu sulfide mineralization, the age determination of the Aguablanca stock became a priority objective due to its important implications on exploration. Casquet et al. (1998, 2001) considered the Aguablanca intrusion as a mafic facies of the Santa Olalla intrusion and obtained by Rb–Sr whole rock geochronology an errorchrone of 359 ± 18 Ma for the SOIC. Montero et al. (2000) obtained by the method Pb–Pb Kober an age of 332 ± 3 Ma for the Santa Olalla tonalite. Almost simultaneously, Romeo et al. (2004) and Tornos et al. (2004b) carried out isotopic dating of the Aguablanca stock by two different geochronological methods. Tornos et al. (2004b) obtained an age of 338 ± 3 Ma by ^{40}Ar–^{39}Ar on primary phlogopite in gabbronorites. Meanwhile, Romeo et al. (2004) dated a gabbronorite mingled with a felsic hybrid rock in a zone exposed in the bed of the Ribera de Cala River (Fig. 3.2e), near the contact with the Santa Olalla intrusion, yielding an age of 341 ± 1.5 Ma (U–Pb ID-TIMS on zircons). Field, petrological and textural evidence suggest that the age of this gabbronorite represents the age of the Aguablanca stock, so the age of this gabbronorite was interpreted as the age of the Aguablanca stock. The geochronological work of Romeo et al. (2004) was part of a complete geochronological study including not only the main Santa Olalla and Aguablanca intrusions but also the small Teuler, Garrote and Sultana bodies. All samples yielded ages clustering around 340 ± 3 Ma (U–Pb ID-TIMS on zircons: Santa Olalla tonalite, 341.5 ± 3 Ma; Garrote granite, 339 ± 3 Ma; Teuler granite, 338 ± 2 Ma; and Sultana tonalite, 341 ± 3 Ma, Romeo et al. 2006b) indicating that all these intrusions formed part of the same magmatic event. Later, Ordoñez-Casado et al. (2008) obtained a slightly older age for the Aguablanca gabbro of 344 ± 2.1 Ma (SHRIMP U–Pb on zircons), and Mathur et àl. (2008) obtained a Re–Os isochron with an age of 383 ± 59 Ma for disseminated sulfides. In summary, field, geochronological, structural, geochemical and isotope evidence indicate that Aguablanca and Santa Olalla intrusions formed part of a same magmatic event developed in Early Carboniferous time and that the Aguablanca mineralization took place during the Variscan orogeny.

3.5 Structure and Emplacement of the SOIC

The SOIC is structurally located between the Cherneca Fault to the north, and the Zufre Fault to the south (Romeo et al. 2006a, 2008) (Fig. 3.3). The Cherneca Fault is a SW-verging structure trending parallel to the general Variscan direction in this zone (N120°) with a reverse and sinistral kinematics, whereas the Zufre Fault is a late N80° sinistral strike-slip fault that cuts off the igneous complex to the south. The SOIC was intensely deformed during its magmatic emplacement and crystallization but does not show any evidence of significant subsolidus deformation. With the aim of knowing in detail the structure and geometry in depth of the SOIC, Romeo et al. (2006a) completed a magmatic foliation map along the entire igneous complex (Fig. 3.8)

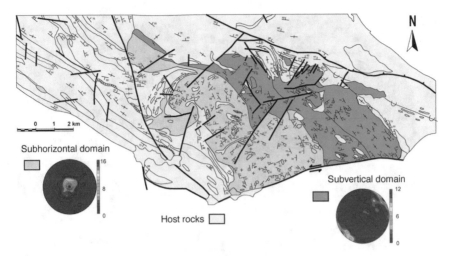

Fig. 3.8 Magmatic foliation map of the Santa Olalla Igneous Complex showing two well-defined different structural domains: NE domain with NW-SE vertical foliations, and SW domain with subhorizontal foliations. Lower-hemisphere equal-area projections of the magmatic foliations from each structural domain are shown (Modified from printed figure by Romeo et al. (2006a), with permission from Elsevier)

and an exhaustive collection of gravity data that led to a 3D reconstruction of the complex (Fig. 3.9). Two different structural domains were defined: (a) NE domain with foliations showing dominantly NW-SE strikes and vertical or high angle dips that is coincident with a thickening of the complex adjacent to the Cherneca Fault (hereafter, subvertical domain), and (b) a thinner SW domain with subhorizontal sheet geometry and predominantly subhorizontal foliations parallel to the upper and lower intrusive contacts (hereafter, subhorizontal domain). Between both domains, a transition zone was also observed, characterized by subvertical foliations super-imposed on the subhorizontal foliations. The Zufre fault, located at the south of the SOIC, cuts the magmatic foliations of tonalite providing its post-intrusive age.

The NE subvertical domain is concordant with the trajectories of the host Bodonal-Cala volcano-sedimentary complex and shows a clear parallelism with the adjacent Cherneca fault. The foliation pattern in the NE domain defines rhomboidal geometries with two main trajectories similar to S–C microstructures formed under non-coaxial shear (Romeo et al. 2006a): one with a N130° strike and 70–90° south dip, and another with a N155° strike and vertical dips. Romeo et al. (2006a) suggested that the magmatic foliations of the subvertical domain were caused by the shear associated to the Cherneca fault strain field, so that the sinistral strike-slip movement of the fault would have been the responsible of the rhomboidal foliation patterns. Assuming that the deepest parts of intrusions are most likely to be the entrance ways for magmas, the NE contact of the SOIC was interpreted to be a feeder zone, and the Cherneca fault probably the conduit used for magma ascent. By the way, in the SW subhorizontal domain, the tonalite is thinner and exhibits a subhorizontal sheet geometry that is

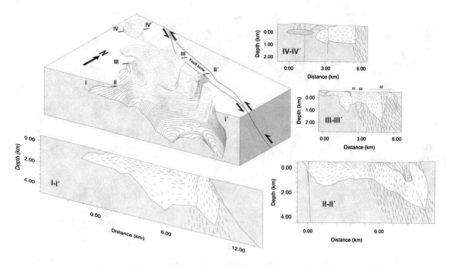

Fig. 3.9 3D structure of the Santa Olalla Igneous Complex obtained from the gravity modelling carried by Romeo et al. (2006a). The 3D image has been reconstructed by the extrapolation of the gravity profiles I–I', II–II', III–III', and IV–IV' (Reprinted from Romeo et al. (2006a), with permission from Elsevier)

not parallel to that of the host rocks exhibiting a subvertical structure. According to Romeo et al. (2006a), these subhorizontal foliations and geometries may be the result of the stress tensor of the Variscan collision being accommodated by depressing the floor and lifting the roof, giving rise to lopolithic geometry at the SW.

The Aguablanca stock belongs to the subvertical domain of the SOIC. The magmatic foliations defined by the preferred orientation of planar crystals of plagioclase (1–4 mm long) are mostly vertical in parallel to the pluton boundaries near to the N, NE and NW contact with the Bodonal-Cala rocks and predominantly subhorizontal in the central part (Fig. 3.10a). In the south, magmatic foliations are linked to the magmatic structure of the surrounding Santa Olalla intrusion with N150° strike and high dip angle towards the NE. The intensity of magmatic foliations is higher toward the NE contact of the Aguablanca intrusion with the Cherneca shear zone. 3D gravity modelling (Romeo 2006; Romeo et al. 2008) has revealed that the Aguablanca stock has inverted drop geometry and that the root for the intrusion, located in the northern margin of the stock adjacent to the Cherneca Fault, has a vertical wedge shape (Fig. 3.10b). This led to these authors to conclude that the Cherneca Fault was probably the magma feeder structure for the Aguablanca intrusion. Crosscutting relationships between Aguablanca stock and the Cherneca Fault (e.g., intrusive rocks cutting the mylonitic foliation in the NE contact; exoskarn produced by Aguablanca deformed by the Cherneca deformation) suggest that the emplacement of Aguablanca took place after the beginning of the deformation associated with the Cherneca Fault. The long axis of the root for Aguablanca intrusion (N65°E) is not parallel to the Cherneca Fault (N115°E) but also seems to be coincident with the orientations

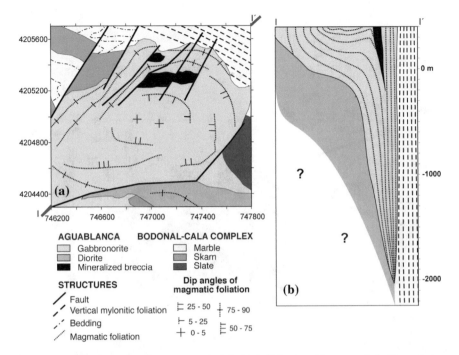

Fig. 3.10 **a** Schematic map of the magmatic foliations of the Aguablanca stock. **b** Cross-section I–I' based on the 3D gravity modelling of Romeo et al. (2008) showing the inverted drop geometry of the Aguablanca stock and the vertical wedge shape of its root adjacent to the Cherneca Fault (Modified from printed figure by Ore Geology Reviews, https://doi.org/10.1016/j.oregeorev.2018. 03.004, Barnes SJ, Piña R, LeVaillant M, Textural development in sulfide-matrix ore breccias in the Aguablanca Ni–Cu deposit, Spain, revealed by X-ray fluorescence microscopy, 2018, with permission from Elsevier)

expected for tension cracks developed in a sinistral ductile strike-slip shear zone with the strike corresponding to the Cherneca Fault (Fig. 3.11). This led to Romeo et al. (2008) to propose that the Aguablanca stock, and the Ni–Cu mineralized breccia pipes, may be emplaced along successive opening hundred-metre-scale tensional cracks formed within the strain field of the Cherneca Fault. Once the magma reached its current erosion level ascending along tensional cracks, it expanded towards SW adopting subhorizontal foliations and its inverted drop geometry (Fig. 3.10b).

In summary, the emplacement and structural evolution for the SOIC can be seen in three stages (Romeo et al. 2008): (1) Magma ascent by the trace of the Cherneca fault during its syntectonic sinistral movement; (2) Once magma reached its present level of emplacement, it propagated toward SW as a horizontal sheet-like intrusion; and (3) After emplacement, the sinistral motion of the Cherneca fault provoked a tectonic stress field in the NE half of the complex giving rise to the subvertical foliation domain.

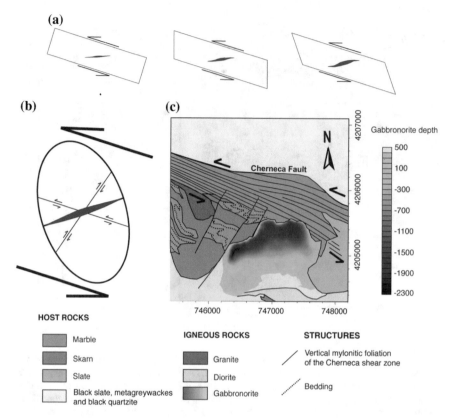

Fig. 3.11 Emplacement model of the Aguablanca stock along open tensional cracks developed during the sinistral displacement of the Cherneca ductile shear zone (Romeo et al. 2008). **a** Formation of tensional cracks in a sinistral shear zone. **b** Strain ellipse deduced for the Cherneca shear zone indicating the expected orientation of tensional cracks. **c** Geological map of the Aguablanca stock and the Cherneca shear zone indicating the orientation of the root of the Aguablanca stock (darker area) parallel to the expected orientation of tensional cracks in the Cherneca shear zone (Reprinted from Romeo et al. (2008), with permission from Cambridge University Press)

References

Apalategui O, Contreras F, Eguiluz L (1990) Santa Olalla de Cala map report. Instituto Geológico y Minero de España (IGME), Mapa Geológico de España MAGNA (1:50000), Sheet 918

Bateman R, Martin MP, Castro A (1992) Mixing of cordierite granitoid and pyroxene gabbro, and fractionation, in the Santa Olalla tonalite (Andalucia). Lithos 28:111–131

Casquet C (1980) Fenómenos de endomorfismo, metamorfismo y metasomatismo en los mármoles de la Rivera de Cala (Sierra Morena). PhD thesis, Universidad Complutense de Madrid, Spain, p 290

Casquet C, Eguiluz L, Galindo C, Tornos F, Velasco F (1998) The Aguablanca Cu–Ni–(PGE) intraplutonic ore deposit (Extremadura, Spain). Isotope (Sr, Nd, S) constraints on the source and evolution of magmas and sulfides. Geogaceta 24:71–74

Casquet C, Galindo C, Tornos F, Velasco F, Canales A (2001) The Aguablanca Cu–Ni ore deposit (Extremadura, Spain), a case of synorogenic orthomagmatic mineralization: age and isotope composition of magmas (Sr, Nd) and ore (S). Ore Geol Rev 18:237–250

Eguiluz L (1988) Petrogénesis de rocas ígneas y metamórficas en el antiforme Burguillos–Monesterio, Macizo Ibérico meridional. PhD thesis, Universidad del País Vasco, Spain, p 694

Eguiluz L, Carracedo M, Apalategui O (1989) Stock de Santa Olalla de Cala (Zona de Ossa-Morena, España). Stvdia Geológica Salmanticensia 4:145–157

Mathur R, Tornos F, Barra F (2008) The Aguablanca Ni–Cu deposit: a Re–Os isotope study. Int Geol Rev 50:948–958

Montero P, Salman K, Bea F (2000) New data on the geochronology of the Ossa-Morena Zone, Iberian Massif. Basement Tectonics 15:136–138

Ordóñez-Casado B, Martín-Izard A, García-Nieto J (2008) SHRIMP-zircon U–Pb dating of the Ni–Cu–PGE mineralized Aguablanca gabbro and Santa Olalla granodiorite: confirmation of an early carboniferous metallogenic epoch in the Variscan Massif of the Iberian Peninsula. Ore Geol Rev 34:343–353

Pereira MF, Chichorro M, Linnemann U, Eguiluz L, Brandao Silva J (2006) Inherited arc signature in Ediacaran and Early Cambrian basins of the Ossa-Morena Zone (Iberian Massif, Portugal): paleogeographic link with European and North African Cadomian correlatives. Precambrian Res 144:297–315

Piña R (2006) El yacimiento de Ni–Cu–EGP de Aguablanca (Badajoz): Caracterización y modelización metalogenética. PhD thesis, Universidad Complutense de Madrid, Spain, p 254

Piña R, Lunar R, Ortega L, Gervilla F, Alapieti T, Martínez C (2006) Petrology and geochemistry of mafic-ultramafic fragments from the Aguablanca (SW Spain) Ni–Cu ore breccia: Implications for the genesis of the deposit. Econ Geol 101:865–881

Pous J, Muñoz G, Wiebke H, Melgarejo JC, Quesada C (2004) Electromagnetic imaging of Variscan crustal structures in SW Iberia: the role of interconnected graphite. Earth Planet Sci Lett 217:435–450

Romeo I (2006) Estudio estructural, gravimétrico y geocronológico del Complejo Ígneo de Santa Olalla (SO de la Península Ibérica): marco tectónico del yacimiento de Ni–Cu–(EGP) de Aguablanca. PhD thesis, Universidad Complutense de Madrid, Spain, p 228

Romeo I, Lunar R, Capote R, Dunning GR, Piña R, Ortega L (2004) Edades de cristalización U–Pb en circones del complejo ígneo de Santa Olalla de Cala: implicaciones en la edad del yacimiento de Ni–Cu–EGP de Aguablanca (Badajoz). Macla 2:29–30

Romeo I, Capote R, Tejero R, Lunar R, Quesada C (2006a) Magma emplacement in transpression: the Santa Olalla Igneous Complex (Ossa-Morena Zone, SW Iberia). J Struct Geol 28:1821–1834

Romeo I, Lunar R, Capote R, Quesada C, Dunning GR, Piña R, Ortega L (2006b) U/Pb age constraints on Variscan Magmatism and Ni–Cu–PGE metallogeny in the Ossa-Morena Zone (SW Iberia). J Geol Soc 163:837–846

Romeo I, Capote R, Lunar R (2007) Crystallographic preferred orientations and microstructure of a Variscan marble mylonite in the Ossa-Morena Zone (SW Iberia). J Struct Geol 29:1353–1368

Romeo I, Tejero R, Capote R, Lunar R (2008) 3-D gravity modelling of the Aguablanca Stock, tectonic control and emplacement of a Variscan gabbronorite bearing a Ni–Cu–PGE ore, SW Iberia. Geol Mag 145:345–359

Schäfer HJ, Gebauer D, Nägler TF, Eguiluz L (1993) Conventional and ion-microprobe U–Pb dating of detritical zircons of the Tentudia Group (Serie Negra, SW Spain): implications for zircon systematics, stratigraphy, tectonics and the Precambrian/Cambrian boundary. Contrib Mineral Petrol 113:289–299

Tornos F, Casquet C, Galindo C, Velasco F, Canales A (2001) A new style of Ni–Cu mineralization related to magmatic breccia pipes in a transpressional magmatic arc, Aguablanca, Spain. Mineral Deposita 36:700–706

Tornos F, Inverno CMC, Casquet C, Mateus A, Ortiz G, Oliveira V (2004a) The metallogenic evolution of the Ossa-Morena Zone. J Iberian Geol 30:143–181

Tornos F, Iriondo A, Casquet C, Galindo C (2004b) Geocronología Ar–Ar de flogopitas del stock de
    Aguablanca (Badajoz). Implicaciones sobre la edad del plutón y de la mineralización de Ni–(Cu)
    asociada. Geotemas 6:189–192
Velasco F (1976) Mineralogía y metalogenia de las skarns de Santa Olalla (Huelva). PhD thesis,
    Universidad del País Vasco, Spain, p 290
Velasco F, Amigó JM (1981) Mineralogy and origin of the skarn from Cala (Huelva, Spain). Econ
    Geol 76:719–727

Chapter 4
The Aguablanca Ni–Cu–(PGE) Sulfide Deposit

4.1 Spatial Distribution of Orebodies: The Mineralized Breccia

The Aguablanca Ni–Cu sulfide mineralization occurs in form of a poorly exposed subvertical mineralized breccia in the northern part of the Aguablanca stock, near to the sedimentary host rocks. The mineralized breccia consists of unmineralized (or very weakly mineralized) mafic-ultramafic fragments embedded in a variably mineralized matrix (Fig. 4.1). The breccia forms a subvertical funnel about 250–300 m wide (N–S), about 600 m long (E–W) and a dip of 70°–80°N. From 600 m depth, the dip of the breccia shallows to 60°S (Fig. 4.2). At this point, the mineralized breccia is apparently truncated by a subvertical ultramafic body. The downward continuity of the breccia is unknown because drilling and geophysical surveying has been indefinitely stopped on closure of the mine at the beginning of 2016.

Within the mineralized breccia, the sulfide mineralization is concentrated in two northward dipping and E-W-trending subvertical bodies characterized by Ni grades higher than 0.08 wt%. Because these orebodies have been outlined from Ni grades, they do not really represent geological or mineralogical contacts, but also uniquely represent the strongly mineralized parts within breccia. The north orebody is the smallest one, occurs very close to the host rocks and reaches a depth of 160 m, whereas the south orebody extends to more than 650 m deep. Both orebodies are locally truncated by NE-oriented strike-slip faults. Out of these orebodies, the sulfide content is significantly lower but locally some zones contain more than 0.08 wt% Ni grades.

The breccia comprises a variably mineralized matrix hosting unmineralized mafic-ultramafic igneous fragments. Two main types of ore are distinguished based on sulfide abundance: semi-massive and disseminated ores. The semi-massive ore consists of up to 85 modal % sulfides, commonly between 20 and 70 modal %. These sulfides poikilitically enclose euhedral to subhedral crystals of olivine, pyroxene and/or plagioclase (Fig. 4.3a, b). The disseminated ore is volumetrically more abundant than

© The Author(s) 2019
R. Piña, *The Ni-Cu-(PGE) Aguablanca Ore Deposit (SW Spain)*,
SpringerBriefs in World Mineral Deposits, https://doi.org/10.1007/978-3-319-93154-8_4

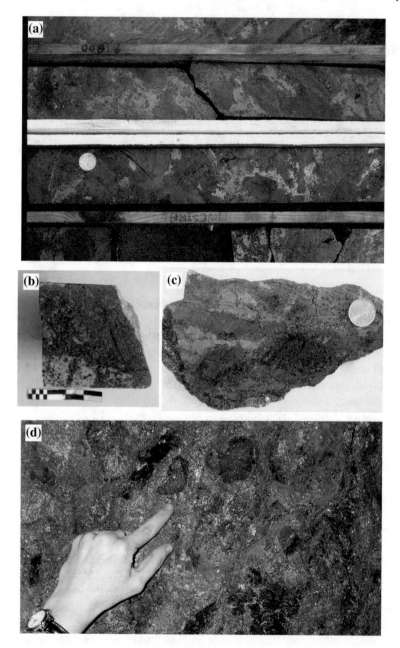

Fig. 4.1 Photographs of drill cores (**a**, **b**), hand sample (**c**), and field taken in the exploration gallery (**d**) illustrating the ore-bearing breccia consisting of unmineralized mafic-ultramafic igneous and metasedimentary fragments hosted by variably mineralized rocks

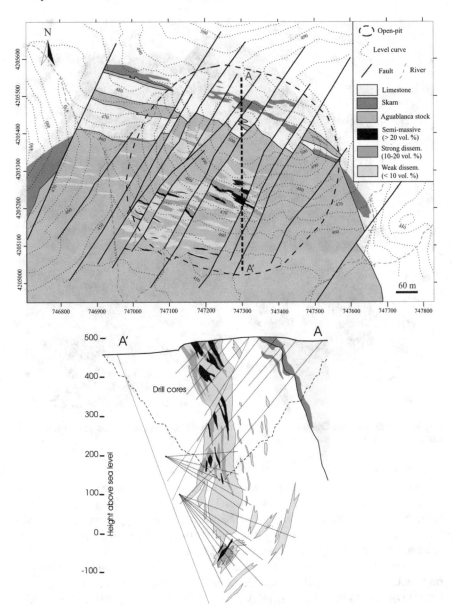

Fig. 4.2 Simplified geological map (**a**) and cross-section (**b**) of the Aguablanca ore-bearing breccia at depth 450 as interpreted by drill cores showing the disposition of ore-types (Modified from printed figure Barnes et al. (2018), with permission from Elsevier)

the semi-massive ore and consists of sulfides (<20 modal %) situated interstitially between silicate minerals in a gabbronorite rock (Fig. 4.3c, d). The mineralized brec-

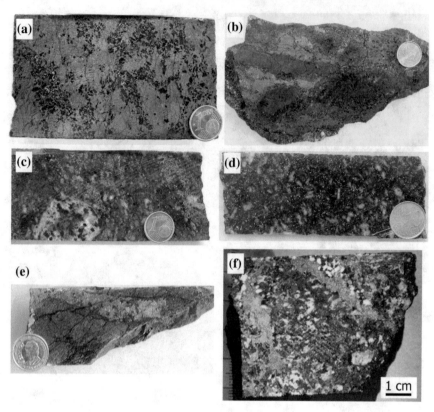

Fig. 4.3 Types of mineralization in the Aguablanca Ni–Cu deposit: semi-massive (**a**, **b**), disseminated (**c**, **d**) and chalcopyrite veined (**e**, **f**) ores

cia is concentrically zoned; in general, the semi-massive ore occurs in the core of the breccia surrounded by the disseminated ore (Fig. 4.2), which grades outwards laterally without systematic changes in the silicate mineralogy to sulfide-free gabbronorite. Minor chalcopyrite-rich veins occur crosscutting both semi-massive and disseminated ore throughout the deposit (Fig. 4.3e, f).

Unmineralized mafic-ultramafic and country-rock (mainly, calc-silicate hornfels) fragments are preferentially concentrated in the semi-massive ore, but they also occur in the disseminated ore and even in sulfide-free gabbronorite (Fig. 4.4). Mafic-ultramafic igneous fragments comprise different cumulate-textured rock types (Piña et al. 2006): peridotite (dunite, harzburgite and werhlite), pyroxenite (orthopyroxenite and clinopyroxenite), gabbro (gabbro, hornblende gabbro and gabbronorite) and anorthosite. Gabbro *s.s.* (clinopyroxene + plagioclase) is by far the most abundant fragment type. In general, country-rock fragments have sharper contacts with the mineralized matrix than mafic-ultramafic fragments. Both fragment-types have commonly subangular to rounded shapes, and a size ranging from few centimetres to 10 cm. Synchrotron-based microbeam XRF mapping has revealed that mafic-

Fig. 4.4 Unmineralized
mafic fragment hosted by
Aguablanca gabbronorite
rock

ultramafic fragments surrounded by sulfides show some degree of disaggregation along the original grain boundaries, whereas the country-rock fragments show only minor reaction rims (Barnes et al. 2018). Locally, pyroxene and plagioclase laths of the gabbronorite matrix appear to have nucleated on the fragments and tabular crystals of plagioclase show planar alignment parallel to the fragment boundaries suggesting some orientation during flow of the sulfide magma. Fragments are commonly unmineralized, although some host very minor disseminations interstitial to primary silicates or in association with secondary minerals such as epidote or actinolite. Minor chalcopyrite veinlets cut both fragments and host rocks. Based on the cumulate textures, the wide range of rock types from ultramafic to mafic, and a range of Mg-numbers in the primary ferromagnesian silicates, the mafic-ultramafic igneous fragments have been interpreted to be derived from previously crystallized rocks representing different stages of cumulate formation in an underlying differentiating magma chamber (Piña et al. 2006).

4.2 Gossan

In the shallow parts of the Aguablanca mineralized breccia there is a supergene oxidation profile ~10 m thick (Fig. 4.5). This gossan has been divided in an "upper unit" dominated by massive goethite, and a less oxidized "lower unit" where primary sulfides are relatively well preserved (Suárez et al. 2010). The gossan mineralogy comprises goethite, hematite, limonite and minor garnierite, relicts of base metal sulfides, inherited silicates, and a complex assemblage of alteration minerals including actinolite, talc, chlorite and quartz. The study of the gossan carried out by Suárez et al. (2010) revealed that gossan has relatively high PGE contents, 1723 ppb total PGE in the upper gossan and 3418 ppb in the lower gossan, similar to the PGE abundances in the underlying unweathered sulfide mineralization. These

Fig. 4.5 Field (**a**) and hand sample (**b**) photographs of the Aguablanca gossan

authors described a great variety of PGE-bearing minerals in the lower unit of gossan
in close textural relationship with goethite and silicates: Pt- and Pd-bearing oxides,
relics of PGM documented in the unweathered sulfide mineralization, PGE-bearing
Fe ($\pm$ Ni–Cu)–oxides (up to 1.1 wt% Pt and 5.4 wt% Pd), PGE-hydroxides, and
PGE-bearing goethite and hematite. According to these authors, these phases are the
result of in situ alteration and dispersion of PGE during gossan formation.

4.3 Mineralized Matrix

The matrix to the breccias takes two main forms: (a) semi-massive ore, and (b)
disseminated ore. In addition, chalcopyrite veinlets (chalcopyrite-veined ore) occur
crosscutting both fragments and mineralized matrix.

4.3.1 Semi-massive Ore

Semi-massive ore preferentially occupies the inner zones of the breccia, which are
surrounded by disseminated ore. Texturally, semi-massive ore is characterized by
euhedral to subhedral (up to 0.5 cm across) silicates enclosed by an assemblage of sulfides composed of pyrrhotite and pentlandite with minor amounts of chalcopyrite and
pyrite. The modal proportion of sulfides typically ranges from 45 to 70% but locally
reaches modal proportions as high as 85%. Silicates comprise variable proportions
of orthopyroxene (<48%), plagioclase (<30%), clinopyroxene (<27%) and minor
amounts of hornblende (<9%), phlogopite (<3%) and olivine (<2%) (Fig. 4.6a, b).
This primary silicate assemblage is variably replaced to sericite, serpentine, bastite,
talc, actinolite, chlorite, calcite and clay minerals.

Orthopyroxene occurs in form of euhedral crystals (0.3–4 mm, typically <3 mm)
variably altered to bastite along fractures and cleavage planes (Fig. 4.6c), talc, acti-

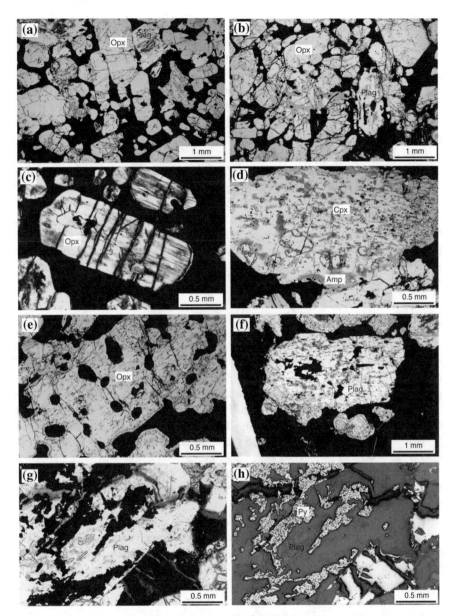

Fig. 4.6 Representative transmitted-light photomicrographs of the semi-massive ore. **a, b** Sulfides enveloping early-formed pyroxene and plagioclase crystals. **c** Euhedral crystal of orthopyroxene partially altered to bastite within sulfides. **d** Clinopyroxene patchy replaced to actinolite. **e** Orthopyroxene hosting subrounded inclusions of magmatic sulfides. **f** Euhedral plagioclase enclosing sulfides. **g, h** Plagioclase partially replaced by secondary pyrite (**g** transmitted-light; **h** reflected-light). Abbreviations: Opx orthopyroxene, Plg plagioclase, Amp amphibole, Py pyrite

nolite and calcite. In places, it shows thin clinopyroxene exsolution lamellae along
(100) planes. Clinopyroxene forms subhedral grains (0.5–5 mm, typically <3.5 mm)
commonly replaced by actinolite and phlogopite (Fig. 4.6d). Some clinopyroxene
grains show complex oscillatory zoning with respect to the Cr content, consisting of Cr-poor cores, concentric Cr-rich zones, and Cr-poor outermost rims (Barnes
et al. 2018). Some clinopyroxene crystals contain orthopyroxene inclusions, suggesting that orthopyroxene crystallized early, and small subrounded sulfide inclusions
(Fig. 4.6e) pointing the existence of an immiscible sulfide liquid before and/or during
its crystallization. Tabular plagioclase crystals are irregularly altered to sericite and,
at a lesser extent, chlorite. At the most strongly mineralized zones, plagioclase hosts
inclusions of magmatic sulfides (mainly, pyrrhotite and chalcopyrite, Fig. 4.6f) and
is partially replaced by pyrite (Fig. 4.6g, h). Olivine is a minor cumulus silicate,
forming subrounded grains (<0.6 mm) totally replaced by serpentine. Phlogopite,
hornblende and minor quartz comprise interstitial phases located between pyroxene,
plagioclase and sulfides. Hornblende and phlogopite usually have sharp contacts with
plagioclase but irregular with pyroxenes, suggesting formation by reaction between
pyroxene and silicate liquid.

Ore minerals

The ore assemblage is typical of magmatic Ni–Cu–Fe sulfide mineralizations, consisting of pyrrhotite ($Fe_{1-x}S$), pentlandite [(Fe, Ni)$_9S_8$], chalcopyrite ($FeCuS_2$) and
minor pyrite (FeS_2). Magnetite (Fe_3O_4), minor amounts of violarite ($FeNi_2S_4$) and
marcasite (FeS_2) replacing pentlandite and pyrrhotite, respectively, and a complex assemblage of platinum-group minerals (PGM), mostly Pd–Pt–Ni bismuthotellurides, complete the ore assemblage. Pyrrhotite is, by far, the most abundant
sulfide in the semi-massive ore, occurring as large (1–2 mm) anhedral twinned
crystals in proportions higher than 55 modal % (Fig. 4.7a–d). Pentlandite (10–35
modal %) occurs as polycrystalline, chain-like aggregates surrounding pyrrhotite
(Fig. 4.7b, c), formed by relative high-temperature grain boundary exsolution, and
as small exsolution flames along grain boundaries and fractures within pyrrhotite
(Fig. 4.7d). Chalcopyrite (<10 modal %) forms anhedral grains of variable size randomly distributed between pyrrhotite and pentlandite. Magnetite (up to 2 modal %)
occurs as individual euhedral crystals within sulfides (mostly, pyrrhotite) and in the
contact between sulfides and silicates (Fig. 4.7a–c). The euhedral morphology of
magnetite suggests early crystallization. In the most altered samples, chalcopyrite
and, at a lesser extent, pyrrhotite, are located along cleavage planes of actinolite and
chlorite (Fig. 4.7e, f), suggesting some sulfide remobilization during alteration.

Pyrite is a common sulfide phase in the Aguablanca semi-massive ore. Typically,
its modal abundance oscillates between 5 and 15 modal %, and it can reach up to
20–25 modal % in areas with intense fracturing and alteration. Similarly, the deepest
parts of the mineralized breccia seem to have even higher modal abundance of pyrite,
with contents up to 40–50 modal %. In this deep part of orebodies, pyrite occurs in
form of large euhedral grains with octahedral morphologies. In general, four main
textural types of pyrite are recognized in Aguablanca (Fig. 4.8) (Ortega et al. 2004;
Piña et al. 2013a): (a) large idiomorphic-subidiomorphic crystals (1–5 mm in size)

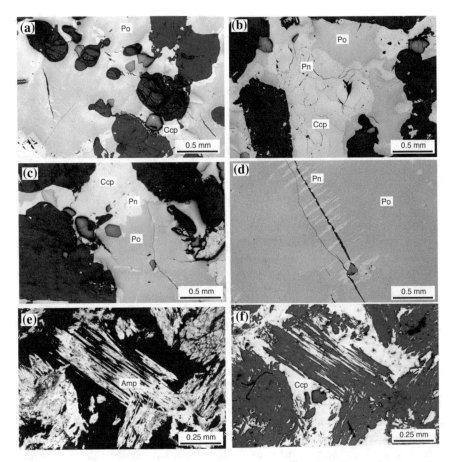

Fig. 4.7 Reflected-light photomicrographs of sulfide textures in the semi-massive ore of the Aguablanca Ni–Cu ore deposit. **a–c** Typical sulfide assemblage formed by pyrrhotite, pentlandite and chalcopyrite with minor amounts of magnetite. **d** Pentlandite flames within pyrrhotite. **e, f** Chalcopyrite occurring along cleavage planes of actinolite (**e**, transmitted light). Abbreviations: Po pyrrhotite, Pn pentlandite, Ccp chalcopyrite, Amp amphibole

within pyrrhotite; (b) ribbon-like crystals (0.5–3 mm long) hosted by pyrrhotite; (c) small subidiomorphic single crystals or aggregates (up to 600 μm) associated with pyrrhotite and chalcopyrite; and (d) irregular crystals partially or totally replacing plagioclase. Ortega et al. (2004) also identified pyrite filling late fractures that crosscut earlier aggregates of ribbon-like pyrite.

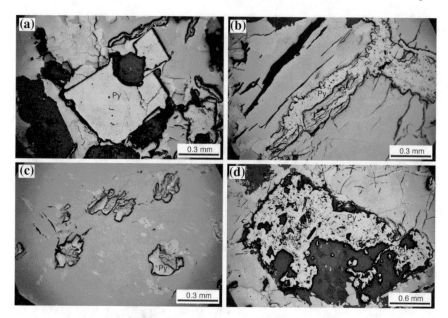

Fig. 4.8 Reflected-light optical photomicrographs showing the different textural types of pyrite recognized in semi-massive ore from the Aguablanca Ni–Cu sulfide ore

4.3.2   Disseminated Ore

In the disseminated ore, sulfides comprising less than 20 vol.% of the rock occur interstitially between primary silicate minerals with sharp and well-defined contacts (Fig. 4.9a, b). Gabbronorite is, by far, the most common lithology hosting dissemi-nated ore, with norite, gabbro and amphibole-rich pyroxenite as other minor igneous rocks. These rocks show cumulate igneous textures formed by variable proportions of cumulus crystals of orthopyroxene (19–56 vol.%), clinopyroxene (<22 vol.%) and plagioclase (<52 vol.%), and intercumulus amphibole (<46 vol.%), phlogopite (<14 vol.%) and minor quartz. Actinolite, sericite, epidote, chlorite, carbonates, talc, serpentine and clay minerals are common secondary phases. Despite locally intense alteration, magmatic textures are usually well preserved and primary silicates are recognized from their morphologies. Chemically, pyroxene and plagioclase of min-eralized gabbronorite are quite similar to these silicates from the unmineralized Aguablanca intrusion as is mainly indicated by the similarity in the Mg and An numbers, respectively. However, the compositions of pyroxene and plagioclase are slightly more primitive in semi-massive ore relative to disseminated ore-bearing gabbronorite.

Ore minerals

Sulfides have typical interstitial disseminated textures (Barnes et al. 2017) in the form of variably sized (from few millimetres to 1–2 cm) polymetallic aggregates between

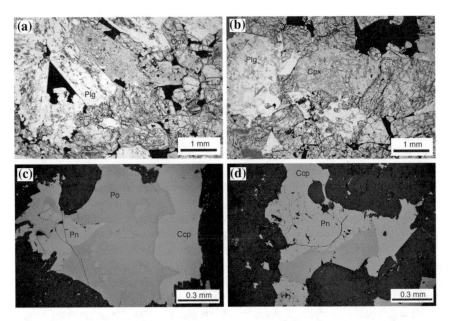

Fig. 4.9 Representative transmitted-light (**a, b**) and reflective light (**c, d**) photomicrographs of disseminated sulfides. **a, b** Gabbronorite hosting disseminated sulfides with sharp and well-defined grain boundaries with sulfides. **c, d** Disseminated sulfides composed of variable amounts of pyrrhotite, pentlandite and chalcopyrite. Abbreviations: Opx orthopyroxene, Cpx clinopyroxene, Plg plagioclase, Po pyrrhotite, Pn pentlandite, Ccp chalcopyrite, Amp amphibole

silicates with well-defined boundaries (Fig. 4.9c, d). The sulfide assemblage is quite similar to that of the semi-massive ore although there are differences in the modal proportions. The most important is that the disseminated ore contains greater proportions of chalcopyrite than the semi-massive ore. Pyrrhotite (22–68 vol.% total ore minerals) is equally the most abundant sulfide but in the disseminated ore chalcopyrite (12–58 vol.%) predominates over pentlandite (up to 20 vol.%) with chalcopyrite/pentlandite ratios ranging typically from 0.06 to 0.64. Another significant difference is that pyrite and magnetite are much less frequent in the disseminated ore. Pyrite is generally absent and, when present, occurs in contents < 5 vol.%.

4.3.3 Chalcopyrite Veins

Chalcopyrite veins represent a very minor ore-type in Aguablanca, comprising less than 5% of the ore present in the deposit. Chalcopyrite veins occur crosscutting both semi-massive and disseminated matrix and mafic-ultramafic fragments. Widths range from a few millimetres to about 2 cm and they extend up to several centimetres in length. They are made up of massive chalcopyrite, minor amounts of pyrrhotite and pentlandite, and traces of argentopentlandite located within chalcopyrite. Pentlandite has more Co (>2 wt%) and less Fe (<27.3 wt%) and pyrrhotite more Ni and Co (~1 and 0.1 wt%, respectively) than in semi-massive and disseminated ores.

4.4 PGE Mineralogy

The sulfide mineralization of Aguablanca includes a relatively abundant assemblage of platinum-group minerals (PGM) (Ortega et al. 2004; Piña et al. 2008; Peralta 2010; Pérez Torrente 2016). Platinum group-minerals are present in all ore-types, although they are notably more abundant in the semi-massive ore (~70%) than in the chalcopyrite veined (~20%) and disseminated (~10%) ores. Most PGM are spatially associated with base-metal sulfides. They occur preferentially included in sulfides (>70%), along sulfide-silicate (~10%) and sulfide-sulfide (~5%) grain boundaries, and only few of them are included in silicates (<10%). Most PGM occur as single grains with sizes generally lower than 10–15 μm, but they occasionally reach up to 50 μm size. More than 90% of the PGM grains are (Pd, Ni, Pt)-bismuthotellurides with the rest being sperrylite (PtAs$_2$), irarsite (Ir, Ru, Rh, Pt)AsS and froodite (PdBi$_2$). Pd-bearing PGM largely predominate over Pt-bearing PGM (Fig. 4.10). (Pd, Ni, Pt)-bismuthotellurides consist, in a decreasing order of abundance, of merenskyite (PdTe$_2$), palladian melonite (NiTe$_2$), michenerite (PdBiTe) and moncheite (PtTe$_2$). Toward the deepest parts of the mineralized breccia, the relative abundance of Pt-bearing PGMs (particularly, moncheite and sperrylite) slightly increases although Pd-bearing PGM still predominates.

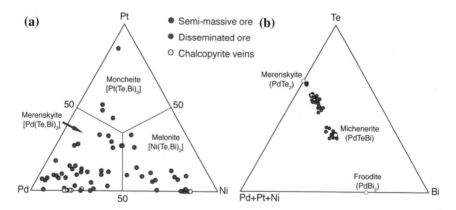

Fig. 4.10 a Modal proportions of merenskyite, melonite and moncheite from the Aguablanca sulfide ores plotted in the PdTe$_2$–NiTe$_2$–PtTe$_2$ end-member ternary diagram. **b** Pd (+Pt, Ni)–Te–Bi (+Sb) ternary diagram showing the compositional variation in at. % of merenskyite and michenerite from the ore-bearing breccia (Modified from printed figure Piña et al. (2008), with permission from Springer)

Merenskyite, Pd (–Ni, Pt)Te$_2$

Merenskyite is the most common PGM in the three ore-types. It forms small (<5–25 μm, but commonly less than 12 μm), rounded to subrounded and elongated grains hosted by pyrrhotite and, less commonly, by pentlandite, chalcopyrite, and sulfide-sulfide and sulfide-silicate interfaces (Fig. 4.11a, b). In the chalcopyrite veinlets, merenskyite occurs within chalcopyrite and along chalcopyrite-silicate grain boundaries. Merenskyite shows wide substitutions of Pd by Ni (0.4–6.5 wt%) and Pt (<17 wt%), and of Te by Bi (4.5–29 wt%) in agreement with the solid-solution series existing between merenskyite-melonite-moncheite (Cabri 2002).

Palladian melonite, Ni (–Pd, Pt)Te$_2$

Palladian melonite preferentially occurs in the semi-massive ore representing about 20% of the total PGM. In general, it forms <10 μm-sized rounded or elongated grains included in pentlandite or attached to pentlandite-silicate and pentlandite-pyrrhotite interfaces (Fig. 4.11c, d). It shows a wide substitution of Ni by Pd (from 3.46 to 12.30 wt% Pd) and of Te by Bi (from 4.43 to 18.67 wt% Bi), and contains variable proportions of Pt (<7.6 wt%).

Michenerite, PdBiTe

Michenerite is present in modal abundances similar to palladian melonite. Its shape and size (normally less than 15 μm) vary considerably, occurring as individual grains with rounded boundaries and, less frequently, with irregular and elongate shapes within pyrrhotite, pentlandite and, at a lesser extent, chalcopyrite (Fig. 4.11e). Palladium contents range from 32 to 36 at. %, Te from 33 to 39 at. % and Bi from 26 to 34 at. %. Traces of Ag and Sb substitute Pd and Bi, respectively.

Moncheite, Pt (–Pd, Ni)Te$_2$

Moncheite is substantially much less frequent than the other bismuthotellurides. It occurs mostly in the semi-massive ore, as small (<10 μm) subrounded inclusions in pentlandite or in the interface pyrrhotite-pentlandite (Fig. 4.11f). Locally, irregular elongated grains occur included in hydrothermal amphibole in the disseminated ore. Its composition reveals extensive substitution of Pt for Pd (up to 10.46 wt%) and Ni (from 1.70 to 12.12 wt%), and Te for Bi (from 11.39 to 25 wt%).

Sperrylite, PtAs$_2$

Sperrylite occurs in the semi-massive and disseminated ores. In general, it forms euhedral crystals enclosed in pyrrhotite, pentlandite and chalcopyrite, with sizes generally larger (15–30 μm) than those for the other PGM. Some sperrylite grains with angular blocky shapes show irregular grain boundaries in contact with chlorite (Fig. 4.12a, b). Sperrylite composition is almost stoichiometric with only minor traces of Fe and S (<1 wt%).

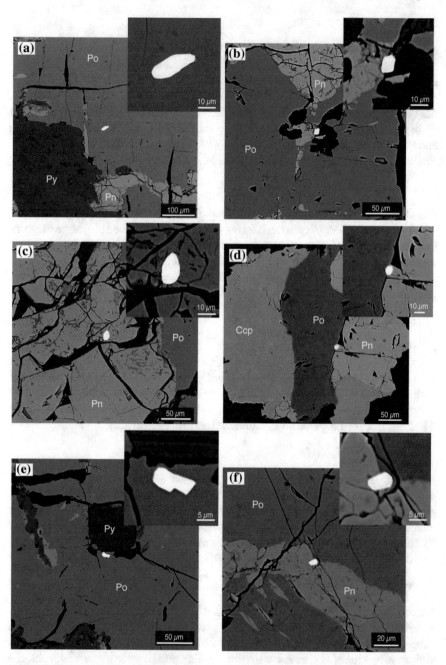

Fig. 4.11 Representative back-scattered electron microprobe images of (Pd, Pt)-bismuthotellurides: merenskyite (**a**, **b**), palladian melonite (**c**, **d**), michenerite (**e**) and moncheite (**f**)

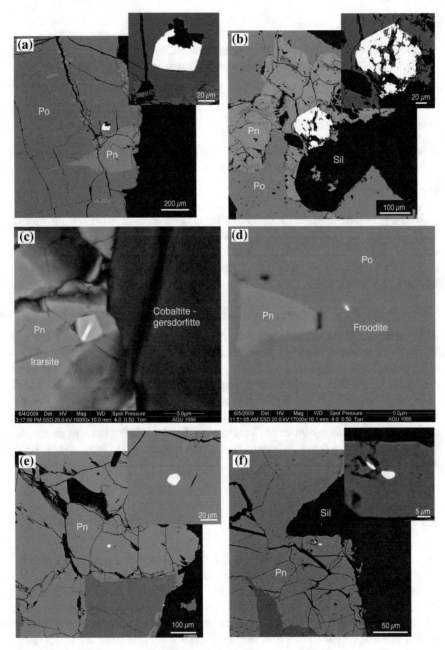

Fig. 4.12 Representative back-scattered electron microprobe images of sperrylite (**a**, **b**), irarsite (**c**), froodite (**d**), tsumoite (**e**) and tellurobismuthite (**f**)

Irarsite, (Ir, Ru, Rh, Pt)AsS

Irarsite preferentially occurs in the deepest parts of semi-massive body, being very scarce in the shallowest zones. It forms tiny (<2 μm) euhedral inclusions enclosed within euhedral grains of cobaltite-gersdorfitte (CoAsS–NiAsS), which are included within pyrrhotite (Fig. 4.12c).

Froodite, $PdBi_2$

It represents a very minor phase in the deposit, occurring as tiny (<2 μm) grains within pyrrhotite (Fig. 4.12d) in semi-massive sulfides.

Cobaltite-gersdorfitte, CoAsS–NiAsS

Cobaltite-gersdorffite occurs both disseminated and semi-massive ore in form of single idiomorphic-subidiomorfic crystals hosted by pyrrhotite. Their Ni, Co and Fe contents vary from 11.5 to 16.6, 13.4 to 17.9, and 3.6 to 8.3 wt%, respectively, and contain traces of Pd (up to 0.64 wt%), Pt (up to 0.96 wt%), Ir (up to 1.8 wt%) and Rh (up to 1.7 wt%).

Other related phases

Along with the PGM assemblage, Piña (2006) also identified a series of Bi- and Te-bearing minerals texturally related to sulfides, including tsumoite (BiTe), telluro-bismuthite (Bi_2Te_3), bismuthinite (Bi_2S_3) and tetradimite (Bi_2Te_2S). Most of these grains are <5 μm in size and occur as inclusions within pyrrhotite and pentlandite from semi-massive and disseminated ore (Fig. 4.12e, f). This study also revealed the presence of argentopentlandite [$Ag(Fe, Ni)_8S_8$] as subhedral inclusions within chalcopyrite in chalcopyrite veins. This mineral has been interpreted as an exsolution product of chalcopyrite in several Ni–Cu occurrences (e.g., Gervilla et al. 1998; Szentpéteri et al. 2002).

4.5 Ore Geochemistry

The average contents of Ni, Cu, PGE and Au of different ore types at Aguablanca are shown in Table 4.1 (both raw concentrations and contents recalculated to 100% sulfides). Data are from Piña et al. (2008) and Peralta (2010). In semi-massive ores, Ni (1.8–6.4 wt%) commonly exceeds Cu (0.2–3.7 wt%) with Ni/Cu ratios averaging 7.6. This trend is reversed in the disseminated ores, where Cu contents (0.5–4 wt%) exceed those of Ni (0.4–1.2 wt%), with Ni/Cu ratios varying from 0.1 to 1.3 (average 0.7), in agreement with the predominance of chalcopyrite over pentlandite in the disseminated sulfides. Copper content in the chalcopyrite veinlets is high (up to 10.62 wt%), with Ni remaining below 1 wt%. Nickel and S are positively correlated (Fig. 4.13a), but Cu shows no correlation with S (Fig. 4.13b). Cobalt ranges from 96 to 2480 ppm and exhibits a good positive correlation with S and Ni (ρ,

Table 4.1 Average Ni, Cu, PGE and Au contents (also in 100% sulfides) of the different ore-types of the Aguablanca sulfide deposit [Data are from Piña et al. (2008) and Peralta (2010)]

Ore-type (Number of samples)	Ni (wt%)	Cu (wt%)	Os (ppb)	Ir (ppb)	Ru (ppb)	Rh (ppb)	Pt (ppb)	Pd (ppb)	Au (ppb)
Raw contents									
SM ore (21)	4.25	1.18	33	82	53	95	700	850	124
D ore (16)	0.63	1.10	3	9	7	11	318	262	208
Ccp veins (3)	0.92	5.66	7	16	12	20	311	1167	400
100% sulfides									
SM ore	8.67	2.71	64	153	103	180	1604	1817	263
D ore	6.77	10.74	37	144	89	118	4910	3022	1992
Ccp veins	4.74	19.71	27	70	51	94	1455	3786	1853

SM, semi-massive; D disseminated; Ccp chalcopyrite

correlation coefficient, = 0.90 and 0.85, respectively) (Fig. 4.13c, d). Sulfur and Se (7–74 ppm) are well correlated each other ($\rho = 0.97$) (Fig. 4.13e). S/Se ratio varies from 2191 to 4710; these values are within the empirical range of mantle-derived sulfides (Naldrett 1981). Gold is highly variable with contents ranging from 14 to 911 ppb. Disseminated ores have higher Au contents than the semi-massive ones. In the few chalcopyrite veinlets identified, Au abundance ranges from 106 to 833 ppb. Gold is only relatively well correlated with Cu ($\rho = 0.52$) (Fig. 4.13f).

Bulk PGE concentrations range from 291 to 3293 ppb and are well correlated with S abundances (Fig. 4.14a). A group of semi-massive samples showing high S contents but low bulk PGE concentration are characterized by being rich in pyrite and exhibiting negative Pt anomalies in the mantle-normalized patterns. Because pyrite and pyrrhotite have similar PGE contents in solid solution, and pyrite hosts lower amounts of PGM than pyrrhotite (<5% of the total PGM), the enrichment in S of this sample group without accompanying PGE increase is likely due to high modal abundances of pyrite. Platinum and Pd predominate over IPGE (Os, Ir, Ru) and Rh with (Pt+Pd)/(IPGE+Rh) ratios ranging from 1 to 303 (80% of the values 5–60). Disseminated ores tend to have higher (Pt+Pd)/(IPGE+Rh) ratios than semi-massive ores, mainly due to the higher content in IPGE and Rh of semi-massive sulfides. IPGE and Rh show strong correlation with each other (Fig. 4.14b), but no correlation with Pt (<0.21) or Pd (<0.27). Osmium, Ir, Ru and Rh contents exhibit strong positive correlations with S (>0.79) (Fig. 4.14c, d). Palladium correlates relatively well with S ($\rho = 0.55$), but Pt is poorly correlated with S ($\rho = 0.21$) (Fig. 4.14e, f). Palladium and Pt are better positively correlated with Te ($\rho = 0.71$ and 0.54, respectively) (Fig. 4.14g, h).

The 100% sulfide-recalculated, mantle-normalized, PGE, Ni, Cu and Au patterns of the different ore-types are broadly similar, with overall positive slopes (Fig. 4.15). The semi-massive ore is relatively enriched in Ni, Os, Ir, Ru and Rh, and depleted

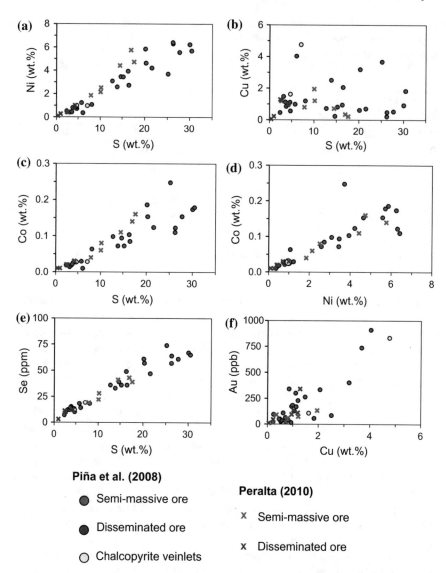

Fig. 4.13 Binary variation diagrams showing the correlations between different elements for the different ore-types present in Aguablanca: **a** Ni versus S; **b** Cu versus S; **c** Co versus S; **d** Co versus Ni; **e** Se versus S; **f** Au versus Cu [Data from Piña et al. (2008) and Peralta (2010)]

in Pd, Pt, Cu and Au with respect to disseminated ore and chalcopyrite veinlets. The Cu and Pd tenors of chalcopyrite veinlets are higher than in disseminated ore. Gold and Pt exhibit a somewhat erratic distribution with significant variations in their abundances within individual ore-types. Semi-massive ore shows lower mantle–normalized Pd/Ir and Cu/Ni ratios than disseminated ore (9.5 and 19.9 versus 22.7 and

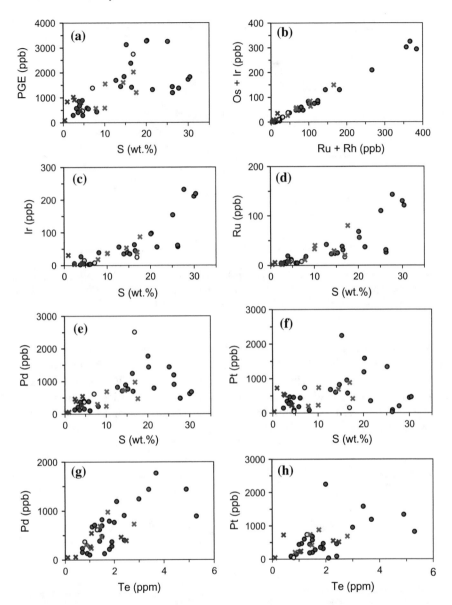

Fig. 4.14 Binary variation diagrams showing the correlations between PGE versus S (**a**), Os+Ir versus Ru+Rh (**b**), Ir versus S (**c**), Ru versus S (**d**), Pd versus S (**e**), Pt versus S (**f**), Pd versus Te (**g**), Pt versus Te (**h**), for the different ore types of the Aguablanca ore deposit (Data are from Piña et al. (2008) and Peralta (2010). Symbols as in Fig. 4.13)

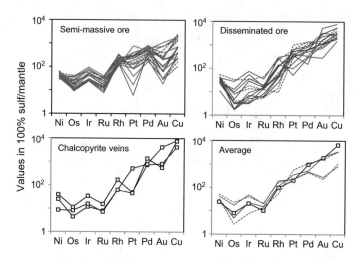

Fig. 4.15 Primitive mantle-normalized metal patterns of the 100% sulfide fraction for the different ore-types in Aguablanca (**a–c**). **d** Mantle-normalized metal patterns for the average values [Data are from Piña et al. (2008) (solid lines) and Peralta (2010) (dotted lines)]

119.3, respectively). The average pattern of chalcopyrite veins shows an even steeper positive slope, with mantle–normalized Pd/Ir and Cu/Ni ratios averaging 48.3 and 296, respectively. In the Ni/Pd versus Cu/Ir and Pd/Ir versus Ni/Cu binary diagrams (Fig. 4.16), there is a gradual decrease in the Ni/Pd and Ni/Cu ratios with the increase of the Cu/Ir and Pd/Ir ratios from the semi-massive to the disseminated ore and chalcopyrite veinlets.

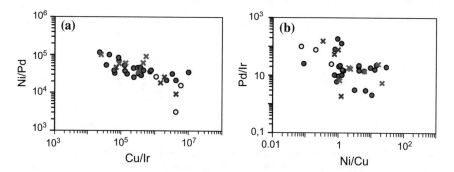

Fig. 4.16 Plots of Ni/Pd versus Cu/Ir (**a**), and Pd/Ir versus Ni/Cu (**b**) for the sulfide ores in the Aguablanca deposit (Data are from Piña et al. (2008) and Peralta (2010). Symbols as in Fig. 4.13)

4.6 In Situ Trace Element Abundances of Sulfides

The concentration of PGE and other trace elements (Au, Ag, Se, Bi, Te, Sb, among others) in pyrrhotite, pentlandite, chalcopyrite and pyrite from the different ore types of Aguablanca deposit was determined by Piña et al. (2012, 2013a) using laser ablation ICP-MS. A brief review of obtained results by these authors is summarized as following.

4.6.1 Pyrrhotite, Pentlandite and Chalcopyrite

Pentlandite is the main sulfide hosting PGE with values ranging from 578 to 7135 ppb, whereas most pyrrhotite and chalcopyrite grains have notably lower PGE concentrations (<400 and 250 ppb PGE, respectively). In general, sulfides from the semi-massive ore have higher PGE contents than those from the disseminated and chalcopyrite-veined ores, mainly due to pyrrhotite and pentlandite from the semi-massive ore contain much more Os, Ir, Ru and Rh than in disseminated ore and chalcopyrite veins (Fig. 4.17).

Pentlandite hosts significant amounts of Pd (466–7087 ppb) as is commonly observed in other Ni–Cu–PGE sulfide deposits (e.g., Merensky Reef, Bushveld Complex, Godel et al. 2007; J-M Reef, Stillwater Complex, Godel and Barnes 2008; Platreef, Bushveld Complex, Holwell and McDonald 2007; Creighton Mine, Sudbury, Dare et al. 2010; Lac des Iles, Duran et al. 2016). Unlike Os, Ir, Ru and Rh, pentlandite from the disseminated and chalcopyrite-veined ores contains more Pd (from 539 to 7087 ppb, and from 3592 to 6770 ppb, respectively) than pentlandite from the semi-massive ore (from 466 to 5218 ppb). Palladium content in pentlandite also seems to depend on the presence of chalcopyrite, because most pentlandite grains in contact with chalcopyrite have higher Pd values than those pentlandites not in contact with chalcopyrite. Flame-textured pentlandites contain ~95% less Pd than coexisting granular pentlandites in a same sample. In pyrrhotite, Pd is below detection limit (~39–46 ppb) in all ore-types, and in chalcopyrite, Pd is more abundant in the disseminated ore (68–407 ppb) and chalcopyrite veined-ore (30–191 ppb) than in the semi-massive ore (up to 82 ppb). Platinum is below the detection limit (~6–10 ppb) in pyrrhotite, pentlandite and chalcopyrite and only some Pd–Pt–Bi–Te microinclusions were identified within pyrrhotite and pentlandite.

Cobalt preferentially occurs in pentlandite with contents ranging from 0.28 to 1.36 wt%. This observation, linked to the excellent positive correlation between Ni and Co in bulk samples, clearly indicates that Co resides within pentlandite in Aguablanca sulfides. For the three ore types, Co in pyrrhotite varies from 50 to 200 ppm, and in chalcopyrite is generally <3 ppm. Selenium is approximately evenly distributed between pyrrhotite, pentlandite and chalcopyrite, being slightly more abundant in the disseminated ore (64–185 ppm) than in the semi-massive (36–80 ppm) and chalcopyrite-veined (43–63 ppm) ores. Bismuth prefer-

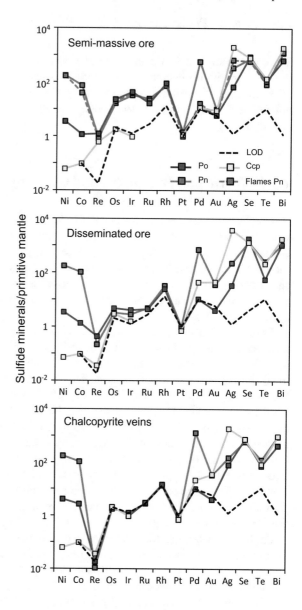

Fig. 4.17 Primitive mantle-normalized element patterns for average values in pyrrhotite, pentlandite (granular and flames) and chalcopyrite from semi-massive (**a**), disseminated (**b**), and chalcopyrite veined (**c**) ores (Data are from Piña et al. (2012). LOD limit of detection)

entially occurs in chalcopyrite with similar concentrations for the three ore types (0.80–9.11 ppm). In pyrrhotite and pentlandite, Bi contents are usually lower (0.16–6.86 ppm and 0.06–14.65 ppm, respectively). The highest Te abundances have been found in pentlandite and chalcopyrite from the disseminated ore (0.7–13.9 ppm in pentlandite and 0.7–5.0 ppm in chalcopyrite). Arsenic concentrations are invariably below the detection limit (~5 ppm). Chalcopyrite hosts most Ag (6–47 ppm) and Cd (3–31 ppm) found in BMS. Silver contents are higher in chalcopyrite from the disseminated ore (12–47 ppm) than in that from the semi-massive one (8–23 ppm). Pentlandite has lower contents (0.4–9 ppm Ag, and 0.01–2.75 ppm Cd) than chalcopyrite, and most pyrrhotite contains less than 1 ppm Ag and 0.2 ppm Cd. Chalcopyrite from the disseminated and chalcopyrite-veined ores hosts from 10 to 170 ppb Au. In the semi-massive ore, the content of Au hosted by chalcopyrite is notably lower (close to the detection limit, ~4–6 ppb). In pentlandite, Au values are approximately equal for the three ore-types (from below the detection limit to 84 ppb) and in pyrrhotite Au is below the detection limit.

4.6.2 Pyrite

A detailed study of PGE and other trace element abundance in the different textural types of pyrite presents in Aguablanca showed the following results (Fig. 4.18) (Piña et al. 2013a): (a) Ribbon-like and small subidiomorphic grains of pyrite occurring within pyrrhotite contain similar Os, Ir, Ru and Rh abundances (100–300 ppb each PGE) to those of host pyrrhotite; (b) pyrite replacing silicates is poor in PGE; (c) large idiomorphic pyrite is much richer in Pt and Rh than any of the other sulfides in Aguablanca; and (d) large idiomorphic crystals of pyrite located within pyrrhotite are compositionally zoned. This zoning consists of alternating Os–Ir–Ru–Rh–As-rich layers and PGE-depleted and Se–Co-rich layers parallel to the grain boundaries (Fig. 4.19a). Some grains further show Pt-rich bands coinciding with high Co values rather than with high IPGE values (Fig. 4.19b). It seems that IPGE competes with Co and Pt for the sites occupied for Fe^{2+}, whereas Se competes with As for the sites occupied for S^2. This type of oscillatory PGE zoning has been also observed in other pyrites of Ni–Cu–PGE ore deposits such as the McCreedy East deposit, Sudbury (Dare et al. 2011) and Lac des Iles, Canada (Duran et al. 2015). This zoning may be the result of boundary layer effect during fluid-assisted solid-state replacement of pyrrhotite by pyrite (Dare et al. 2011). According to this model, at the beginning of

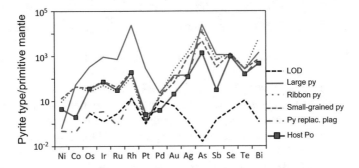

Fig. 4.18 Primitive mantle-normalized element patterns for average values in the different textural types of pyrite and host pyrrhotite (Modified from Piña et al. 2013a)

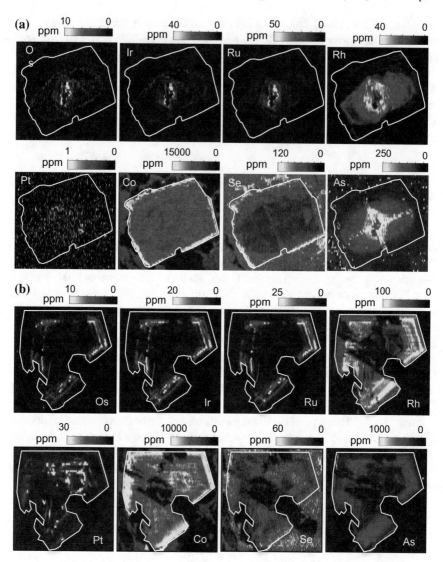

Fig. 4.19 Compositional zoning of idiomorphic pyrite grains in semi-massive ore samples

pyrite growth from pyrrhotite, Os, Ir, Ru, Rh and As would be preferentially incorporated into pyrite leaving the pyrrhotite at the pyrrhotite-pyrite boundary relatively depleted in these elements. If the rate of growth of pyrite were faster than PGE diffusion in pyrrhotite, the next layer of pyrite to grow would form from PGE-depleted pyrrhotite and would incorporate Co and Se instead of PGE and As, respectively. Cafagna and Jugo (2016) have experimentally observed that Co, Os, Ir, Ru, Rh and Re preferentially partition into pyrite from pyrrhotite and these metals show extremely slow diffusion within pyrite. This would explain why pyrite preserves this chemical

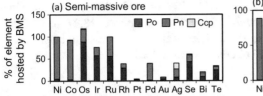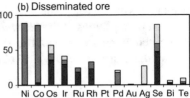

Fig. 4.20 Mass balance of PGE and other chalcophile elements in pyrrhotite, pentlandite and chalcopyrite from the semi-massive (**a**) and disseminated (**b**) ores, plotted as the proportion (%) of element in each sulfide (Modified from printed figure by Piña et al. (2012), with permission from Elsevier)

zonation after it has finished growing. On the other hand, the strong affinity of PGE for As (Piña et al. 2013b) would favour the preferential incorporation of PGE into the As-rich bands.

4.6.3   Mass Balance Calculations

Mass balance calculation (Fig. 4.20; Piña et al. 2012) indicates that most Os, Ir and Ru occur in solid solution within pyrrhotite and, to a lesser extent, pentlandite in the semi-massive ore (~88% Os, ~57% Ir, ~56% Ru for pyrrhotite; ~28% Os, ~19% Ir and ~45% Ru for pentlandite). By contrast, pyrrhotite and pentlandite account for notably lower amounts of these elements in the disseminated ore (~36% Os, ~30% Ir, ~19% Ru for pyrrhotite; ~9% Os, ~6% Ir and ~6% Ru for pentlandite) and chalcopyrite veinlets (~10% Os, ~2% Ir and ~9% Ru in pyrrhotite plus pentlandite). Pentlandite accounts for 15–30% of bulk Pd, whereas chalcopyrite only accounts for 5%. The rest of Pd resides as PGM in the form of Pd-bearing bismuthotellurides. Except for the minor amount of Pt present into pyrite, Pt occurs mostly as PGM (mainly sperrylite, $PtAs_2$). Most Co is hosted by pentlandite (77–96%) and Se occurs mostly within sulfides (~60–96% according the ore-type). The PGE distribution between base metal sulfides and discrete PGM is in agreement with the overall trend found in most Ni–Cu–PGE ore deposits (Barnes and Ripley 2016). In most cases, Os, Ir, Ru and Rh are hosted in large part by pyrrhotite and pentlandite, Pd occurs in significant amounts within pentlandite and Pt is present in form of discrete PGM.

4.7   Sulfur and Lead Isotopes

Sulfur isotope compositions of sulfide minerals in the Aguablanca ores have been studied by Casquet et al. (2001). The $\delta^{34}S$ values of individual pyrrhotite (+7.1 to 7.8‰), pentlandite (+7.4‰) and chalcopyrite (+7.2 to +7.7‰) are very similar each other with a mean value close to +7.4‰. Tornos et al. (2001) determined the lead isotope signatures of the ore with $^{206}Pb/^{204}Pb$ ranging from 18.27 to 18.43 and $^{207}Pb/^{204}Pb$ ranging from 15.61 to 15.65. These values were found to be similar to those of the host rocks and typical of crustal-derived lead.

References

Barnes S-J, Ripley EM (2016) Highly siderophile and strongly chalcophile elements in magmatic ore deposits. Rev Mineral Geochem 81:725–774

Barnes SJ, Piña R, Le Vaillant M (2018) Textural development in sulfide-matrix ore breccias in the Aguablanca Ni–Cu deposit, Spain, revealed by X-ray fluorescence microscopy. Ore Geol Rev 95:849–862

Barnes SJ, Mungall JE, Le Vaillant M, Godel B, Lesher CM, Holwell D, Lightfoot PC, Krivolutskaya N, Wei B (2017) Sulfide-silicate textures in magmatic Ni–Cu–PGE sulfide ore deposits: disseminated and net-textured ores. Am Mineral 102:473–506

Cafagna F, Jugo P (2016) An experimental study on the geochemical behaviour of highly siderophile elements (HSE) and metalloids (As, Se, Sb, Te, Bi) in a mss-iss-pyrite system at 650°C: A possible magmatic origin for Co-HSE-bearing pyrite and the role of metalloid-rich phases in the fractionation. Geochim Cosmochim Acta 178:233–258

Casquet C, Galindo C, Tornos F, Velasco F, Canales A (2001) The Aguablanca Cu–Ni ore deposit (Extremadura, Spain), a case of synorogenic orthomagmatic mineralization: age and isotope composition of magmas (Sr, Nd) and ore (S). Ore Geol Rev 18:237–250

Cabri LJ (2002) The platinum-group minerals: In: Cabri LJ (ed) The geology, mineralogy and mineral beneficiation of platinum-group elements. Canadian Institute of Mining and Metallurgy, Special vol 54, pp 13–129

Dare SAS, Barnes S-J, Prichard HM (2010) The distribution of platinum-group elements (PGE) and other chalcophile elements among sulfides from Creighton Ni–Cu–PGE sulfide deposit, Sudbury, and the origin of palladium in pentlandite. Min Deposita 45:765–793

Dare SAS, Barnes S-J, Prichard HM, Fisher PC (2011) Chalcophile and platinum-group element (PGE) concentrations in the sulfide minerals from the McCreedy East deposit, Sudbury, Canada, and the origin of PGE in pyrite. Min Deposita 46:381–407

Duran CJ, Barnes S-J, Corkery JT (2015) Chalcophile and platinum-group element distribution in pyrites from the sulfide-rich pods of the Lac des Iles Pd deposits, Western Ontario, Canada: implications for post-cumulus re-equilibration of the ore and the use of pyrite compositions in exploration. J Geochem Explor 158:223–242

Duran CJ, Barnes S-J, Corkery JT (2016) Trace element distribution in primary sulfides and Fe–Ti oxides from the sulfide-rich pods of the Lac des Iles Pd deposits, Western Ontario, Canada: constraints on processes controlling the composition of the ore and the use of pentlandite compositions in exploration. J Geochem Explor 166:45–63

Gervilla F, Papunen H, Kojonen K, Johanson B (1998) Platinum-, palladium- and gold-rich arsenide ores from the Kylm/ikoski Ni–Cu deposit (Vammala Nickel Belt, SW Finland). Mineral Petrol 64:163–185

Godel B, Barnes S-J (2008) Platinum-group elements in sulfide minerals and the whole rocks of the J-M Reef (Stillwater Complex): implication for the formation of the reef. Chem Geol 248:272–294

Godel B, Barnes S-J, Maier WD (2007) Platinum-group elements in sulphide minerals, platinum-group minerals, and whole-rocks of the Merensky Reef (Bushveld Complex, South Africa): implications for the formation of the Reef. J Petrol 48:1569–1604

Holwell DA, McDonald I (2007) Distribution of platinum-group elements in the Platreef at Overysel, northern Bushveld Complex: a combined PGM and LA-ICP-MS study. Contrib Mineral Petrol 154:171–190

Naldrett AJ (1981) Nickel sulfide deposits: classification, composition and genesis. Econ Geol 75:628–685

Ortega L, Lunar R, García Palomero F, Moreno T, Martín-Estévez JR, Prichard HM, Fisher PC (2004) The Aguablanca Ni–Cu–PGE deposit, southwestern Iberia: magmatic ore forming processes and retrograde evolution. Can Mineral 42:325–350

Peralta A (2010) Estudio mineralógico y geoquímico del cuerpo profundo del yacimiento de Ni–Cu–EGP de Aguablanca. Universidad de Granada, Spain, MSc, p 52

Pérez Torrente Y (2016) Mineralogía del horizonte 190 rico en pirita del yacimiento de níquel de Aguablanca (Badajoz). MSc. Universidad de Granada, Spain, p 37

Piña R (2006) El yacimiento de Ni–Cu–EGP de Aguablanca (Badajoz): Caracterización y modelización metalogenética. PhD thesis, Universidad Complutense de Madrid, Spain, p 254

Piña R, Lunar R, Ortega L, Gervilla F, Alapieti T, Martínez C (2006) Petrology and geochemistry of mafic-ultramafic fragments from the Aguablanca (SW Spain) Ni–Cu ore breccia: Implications for the genesis of the deposit. Econ Geol 101:865–881

Piña R, Gervilla F, Ortega L, Lunar R (2008) Mineralogy and geochemistry of platinum-group elements in the Aguablanca Ni–Cu deposit (SW Spain). Mineral Petrol 92:259–282

Piña R, Gervilla F, Barnes S-J, Ortega L, Lunar R (2012) Distribution of platinum-group and chalcophile elements in the Aguablanca Ni–Cu sulfide deposit (SW Spain): evidence from a LA-ICP-MS study. Chem Geol 302–303:61–75

Piña R, Gervilla F, Barnes S-J, Ortega L, Lunar R (2013a) Platinum-group elements-bearing pyrite from the Aguablanca Ni–Cu sulphide deposit (SW Spain): a LA-ICP-MS study. Eur J Mineral 25:241–252

Piña R, Gervilla F, Barnes S-J, Ortega L, Lunar R (2013b) Partition coefficients of platinum-group and chalcophile elements between arsenide and sulfide phases as determined in the Beni Bousera Cr–Ni mineralization (North Morocco). Econ Geol 108:935–951

Suárez S, Prichard HM, Velasco F, Fisher PC, McDonald I (2010) Alteration of platinum-group minerals and dispersion of platinum-group elements during progressive weathering of the Aguablanca Ni–Cu deposit, SW Spain. Miner Deposita 45:331–350

Szentpéteri K, Watkinson DH, Molnar F, Jones PC (2002) Platinum-Group Elements-Co-Ni-Fe Sulfarsenides and Mineral Paragenesis in Cu-Ni-Platinum-Group Element Deposits, Copper Cliff North Area, Sudbury, Canada. Econ Geol 97:1459–1470

Tornos F, Casquet C, Galindo C, Velasco F, Canales A (2001) A new style of Ni–Cu mineralization related to magmatic breccia pipes in a transpressional magmatic arc, Aguablanca, Spain. Miner Deposita 36:700–706

Chapter 5
Ore-Forming Processes in the Aguablanca Ore Deposit

The high sulfur content and the low concentrations of PGE (0.47 g/t, evaluation by Río Narcea Recursos, SA) of the Aguablanca mineralization indicate that this deposit belongs, according to the classification of Naldrett (2004), to those magmatic sulfide deposits valuable due to their primary Ni and Cu contents with PGE recovered as by-products.

It is widely accepted that Ni–Cu magmatic sulfide deposits are the result of fractionation and crystallization of immiscible sulfide melts segregated from mantle-derived mafic or ultramafic silicate magmas (Naldrett 2004). Previously, silicate magma must attain the saturation in sulfide liquid, the point where the magma can no longer hold sulfur in solution. At that moment, silicate magma segregates an immiscible sulfide liquid that collects chalcophile elements such as Ni, Cu, Co, Au, Se, and PGE, due to their high partition coefficients between sulfide and silicate melts (Peach et al. 1990; Barnes et al. 1997). Subsequently, the magmatic crystallization and sub-solidus re-equilibration of sulfide melt upon cooling produces the typical mineral association consisting of pyrrhotite + pentlandite + chalcopyrite ± magnetite (Naldrett et al. 1967; Kullerud et al. 1969).

In Aguablanca, textural, mineralogical and geochemical features of sulfide ores clearly indicate a magmatic origin of the mineralization. Supporting evidence for this includes the following: (a) sulfides are not associated with secondary silicates nor are concentrated in intensely-altered zones; by contrast, they are situated interstitially between primary igneous silicates (pyroxene, plagioclase) with well-defined boundaries in the disseminated ore and net-textures in the semi-massive ore; (b) the mineralogical association pyrrhotite + pentlandite + chalcopyrite ± magnetite, considered to be typically the result of sulfide melt crystallization; and (c) the excellent positive correlations between Ni and PGE with S, and the Ir/(Ir + Ru) and Pt/(Pt + Pd) ratios (0.2–0.8 and 0.02–0.9, respectively), typical of magmatic sulfides (0.3–0.7 and 0.3–0.7, respectively, Naldrett et al. 1982).

Ni/Cu and Pd/Ir ratios of the sulfide ores are used to discern the nature of silicate magmas from which sulfides segregated (i.e., Barnes et al. 1987; Keays 1995). Sulfides segregated from ultramafic magmas tend to have Ni/Cu and Pd/Ir ratios

© The Author(s) 2019
R. Piña, *The Ni-Cu-(PGE) Aguablanca Ore Deposit (SW Spain)*,
SpringerBriefs in World Mineral Deposits, https://doi.org/10.1007/978-3-319-93154-8_5

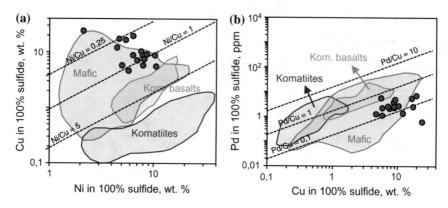

Fig. 5.1 Bulk concentrations of Cu and Ni (**a**) and Pd and Cu (**b**) in 100% sulfides for disseminated sulfides in the Aguablanca Ni–Cu ore deposit. Coloured fields represent the range of metal tenors for sulfides associated with komatiites (purple), mafic-parented intrusions (green) and komatiitic basalts (blue) (Barnes et al. 2017a)

generally higher than 7 and lower than 10, respectively (e.g., Kambalda, Western Australia, Ni/Cu~14 and Pd/Ir~8, Cowden et al., 1986), whereas mafic magmas commonly produce sulfide ores with Ni/Cu and Pd/Ir ratios lower than 2.5 and higher than 10, respectively (e.g., Noril'sk, Russia, Ni/Cu~0.5–1.4 and Pd/Ir~33–160, Naldrett 2004; Jinchuan, China, Ni/Cu~0.3–3.3 and Pd/Ir commonly >15, Su et al. 2008; Kalatongke, China, Ni/Cu~0.25–2.5 and Pd/Ir~4–253, Song and Li 2009). On the basis of Ni/Cu (0.28–1.29) and Pd/Ir (>22) ratios of the disseminated sulfides in Aguablanca (considered formed from unfractionated sulfide melt, Piña et al. 2008), Piña et al. (2012) established a basaltic linkage instead of ultramafic for the parental Aguablanca silicate magmas. This mafic linkage is further supported by the 100% sulfide abundances of Ni, Cu and Pd of disseminated sulfides that plot in the field of mafic magmas of the Cu_{100} versus Ni_{100} and Pd_{100} versus Cu_{100} binary diagrams (Fig. 5.1) (Barnes et al. 2017a).

5.1   Origin of Sulfide Segregation

Several mechanisms can lead to the sulfide saturation of silicate magmas (Ripley and Li 2013 and reference therein): fractional crystallization, magma mixing, assimilation of siliceous country rocks, and addition of externally derived sulfur. Among these processes, sulfur isotope data indicates that the addition of external sulfur is responsible for the formation of many Ni–Cu sulfide ores (Keays and Lightfoot 2010), being currently considered as the most important process triggering sulfide saturation (Ripley and Li 2013). Although there are several mechanisms for adding external sulfur in silicate magmas, the direct melting and assimilation of sulfide-rich crustal material is probably the most efficient mechanism (Robertson et al. 2015).

On the basis of S and Pb isotope date, Casquet et al. (1998) and Tornos et al. (2001) have suggested that the sulfur saturation in Aguablanca likely resulted from the assimilation of graphite- and sulfide-rich black shales belonging to the Late Neo-proterozoic Serie Negra Formation. Data of $\delta^{34}S$ of Aguablanca sulfides (7.4 ± 0.4‰, Casquet et al. 1998) are notably higher than those considered for mantle origin (0 ± 3‰, Ohmoto 1986), whereas the Pb isotope values ($^{206}Pb/^{204}Pb = 18.27-18.43$ and $^{207}Pb/^{204}Pb = 15.61-15.65$, Tornos et al. 2001) are similar to those of the Serie Negra black shales (Tornos and Chiaradia 2004) and typical of rocks with Pb derived from crustal rocks. The Serie Negra Formation dominates the basement along the Olivenza-Monesterio antiform of the Ossa-Morena Zone and locally hosts relatively high sulfur contents (1091 ppm, Nägler 1990; 5238 ppm, Piña 2006). The role of black shales as source of sulfur for sulfide formation has been also documented in other Ni–Cu sulfide deposits (e.g., Duluth complex, USA, Thériault and Barnes 1998; Pechenga Ni deposits, Russia, Barnes et al. 2001). In addition, black shales may be also an important source of TABS (Te, As, Bi, and Sb) because these rocks usually contain between 1 and 3 orders magnitude more of these semimetals than MORB and primitive mantle (e.g., Samalens et al. 2017). Tornos et al. (2006) carried out a mass balance calculation assuming that the $\delta^{34}S$ of the uncontaminated magmas ranged from −3 to +2‰ and that of country rocks from +14 to +21‰ (Tornos and Velasco 2002). They obtained that up to 30% of the sulfur in Aguablanca was of external derivation to the magma. According to Tornos et al. (2006), the assimilation of black shales may be also the responsible of the relatively high Au and Cu contents of the Aguablanca ore, since the Serie Negra locally hosts stratabound mineralizations of chalcopyrite and bornite with Au (Tornos et al. 2004)

Field evidence also supports the involvement of black shales in the sulfide forma-tion. Close to the north boundary of the Aguablanca intrusion, barren gabbronorite hosts locally partially digested xenoliths of black shales. Surrounding these xeno-liths, there is small sulfide patches probably resulted from local segregation of sulfide droplets due to the addition of S and/or SiO_2 from enclaves. Assimilation probably involved partial melting of black shales and release of sulfur to the silicate melt. These sulfides likely represent the same mechanism, but at small scale, that gave rise to the main sulfide segregation in Aguablanca.

S/Se ratios of sulfide ores can be used as petrogenetic indicators for sulfur origin (Queffurus and Barnes 2015). S/Se ratios ranging from 3000 to 4500 are considered as typical of mantle origin (Eckstrand and Hulbert 1987). By contrast, S/Se ratios higher than mantle values are interpreted to be the result of magma contamination with S-rich sedimentary rocks (e.g., Lesher and Burnham 2001; Maier and Barnes 2010), whereas S/Se ratios less than mantle values are interpreted to indicate S-loss during post-crystallization events such as desulfurization by S-undersaturated fluids. Importantly, the use of S/Se ratios by themselves as indicators of sulfur source can lead to wrong interpretations unless the S/Se ratio of the potential contaminant rocks is also known. Aguablanca represents a good example. S/Se ratios in Aguablanca range from 2613 to 4710 (Piña et al. 2006). These values are within the range considered for mantle sulfides and, a priori, do not support a sedimentary origin for sulfur. Nevertheless, if a sedimentary origin for sulfur is considered, as is strongly

Fig. 5.2 Plot of Pt + Pd
(contents recalculated to
100% sulfide) versus S/Se
ratios for the Aguablanca
sulfides. In grey, values for
mantle domain taken from
Eckstrand and Hulbert
(1987)

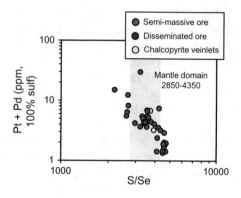

evidenced by sulfur isotope values, the unusually low S/Se ratios may be due to the
S-loss during metamorphism and/or alteration because S is much more mobile than
Se during these processes (Howard 1977; McGoldrick and Keays 1981). However,
the S versus Se binary diagram clearly indicates that S-loss is very improbable (Fig.
4.13e), because some deviation toward the S-loss field would be expected and the
excellent positive S-Se correlation does not deviate at all. Another possibility is that
the low S/Se ratios of sulfide mineralization are due to the initial low S/Se ratios
(from 184 to 3300, Piña 2006) of S-rich black shales from the Serie Negra Formation,
the most probable contaminant. In this scenario, the contamination would not give
as result a high S/Se ratio but also would produce S/Se ratios close to the primary
mantle values. The unusually low S/Se ratios of the black shales may result from
(1) S-loss due to devolatilization reactions related to the replacement of pyrite by
pyrrhotite that typically takes places in these pellitic rocks as has been observed in
other localities such as the Virginia Formation, the country rocks surrounding the
Duluth Complex (Ripley 1990; Thériault and Barnes 1998); (2) high and variable
Se contents of the black shales. In addition, R-factor may play an important role
on the Se concentration (and consequently, S/Se ratios) of the disseminated sulfides
(Queffurus and Barnes 2015). Selenium is a strongly chalcophile element with a
high partition coefficient between sulfide and silicate liquids ($D_{Se}^{sulf/sil} = 1770$, Peach
et al. 1990), so the Se content of sulfide increases (and hence S/Se ratios decrease)
with increasing R-factors. In general, Aguablanca disseminated sulfides have lower
S/Se ratios than semi-massive sulfides, and these ratios show a strong negative
correlation with the Pd + Pt (100% sulfides) abundances (Fig. 5.2). It is probable that
the sulfides forming the disseminated sulfides interacted for a longer time, with a
larger volume of magma, than the semi-massive sulfides, leading to a higher amount
of Se in the sulfide liquid (and consequently decreasing S/Se ratios).

Additional evidence for crustal contamination of the Aguablanca parental magma
is provided by the predominance of orthopyroxene over olivine, the enrichment in
SiO_2, LILE and LREE of the host igneous gabbronorites, and pronounced negative Ta
and Nb and positive Pb anomalies (Fig. 3.7). The extent of crustal contamination in
the parental magmas of Aguablanca rocks can be estimated by using (Th/Yb) versus

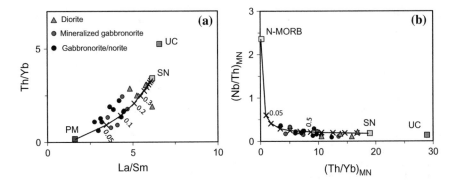

Fig. 5.3 Plot of whole rock (Th/Yb) versus (La/Sm) (**a**), and (Nb/Th)$_{MN}$ versus (Th/Yb)$_{MN}$ (**b**) for the Aguablanca rocks, showing the effect of crustal contamination on trace element ratios. The primitive mantle value (PM) is from McDonough and Sun (1995), and the N-MORB value is from Sun and McDonough (1989). The average value for the country rocks of the Serie Negra Formation is from Pereira et al. (2006). Data for Aguablanca rocks are from Piña (2006), Casquet et al. (2001) and Tornos et al. (2006). The tie lines indicate pure linear mixing between primitive mantle and N-MORB, and Serie Negra Formation, for (**a**) and (**b**), respectively

(La/Sm), and (Nb/Th)$_{PM}$ versus (Th/Yb)$_{PM}$ diagrams (Fig. 5.3). These ratios are sensitive indicators of crustal contamination and their representation along with the ratios of possible contaminants is useful to constrain the degree of contamination. A contamination model is tested in a plot of Th/Yb versus La/Sm (Fig. 5.3a). The average ratios of the Serie Negra Formation (Pereira et al. 2006) are used for the hypothetical contaminant, and the composition of the primitive mantle and N-MORB (according to McDonough and Sun 1995 and Sun and McDonough 1989, respectively) were used as mantle end members. In this plot, the Aguablanca rocks plot close to the field of the country rocks containing between 5 and 30% of crustal component. Figure 5.3b shows that, compared to the N-MORB, the Aguablanca rocks have very high (Th/Yb)$_{PM}$ and low (Nb/Th)$_{PM}$, which is indicative of strong crustal contamination (~10–50%).

Finally, Sr and Nd isotope compositions of the igneous host rocks (Casquet et al. 1998; Tornos et al. 2001) and Re–Os isotope data of primary sulfides (Mathur et al. 2008) also point to contamination processes during the magmatic evolution of Aguablanca. In the first case, the high ^{87}Sr/^{86}Sr$_i$ and low ^{143}Nd/^{144}Nd$_i$ ratios of the Aguablanca igneous rocks fall between the values of unmineralized Variscan plutons of the area and those of the host metasedimentary rocks (Fig. 5.4; Tornos et al. 2001), suggesting Aguablanca magmas assimilated sediments similar to those hosting the Aguablanca stock. On the other hand, Re–Os isotopic signature of the disseminated sulfides is interpreted as the result of simple mixing between the Serie Negra Formation and mantle-derived magmas. Mathur et al. (2008) carried out a numerical modelling concluding that it would be necessary the assimilation of 15–25 wt% of shales by the mantle-derived magma to reproduce the Re–Os isotope values registered in Aguablanca sulfides.

Fig. 5.4 Plot of
^{143}Nd/^{144}Nd versus
^{87}Sr/^{86}Sr ratios for the
Aguablanca, Santa Olalla
and other Variscan rocks and
country host Upper
Proterozoic—Lower
Cambrian rocks showing the
crustal contamination for the
rocks of Aguablanca and
Santa Olalla (Modified from
printed figure by Tornos
et al. (2001), with permission
from Springer)

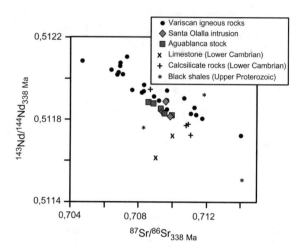

All the evidence provided above indicates that the contamination of Aguablanca magmas with S-rich crustal rocks was the key factor in the formation of sulfide ores, with the black shales of the Serie Negra Formation being the most probable contaminant. It is probable that the contamination of Aguablanca magmas with graphitic- and sulfide-rich black shales not only added significant amounts of S, but also added SiO_2, favouring the sulfide liquid saturation and orthopyroxene crystallization, since the incorporation of silica to mafic magmas decreases their sulfide liquid solubility (Li and Naldrett 1993).

5.2   Timing of Sulfide Segregation

A notable feature of the Aguablanca sulfide ores is that sulfides are intimately related to the mafic-ultramafic fragments, as is demonstrated by the fact that the fragments are particularly abundant in the most strongly mineralized parts of the breccia. Piña et al. (2006) carried out a detailed study of the different types of mafic-ultramafic fragments present in the mineralized breccia and concluded they were probably derived from previously crystallized rocks (probably belonging to a hidden differentiated mafic-ultramafic sequence situated beneath Aguablanca stock) that were brecciated and emplaced as solid (or semisolid) clasts. Tornos et al. (2006) and Piña et al. (2010) suggested that the sulfide melt would have unmixed from the silicate magma before the fractionation of peridotite cumulates in the differentiating complex now preserved as fragments in the breccia. In their discussion about the timing of sulfide segregation, Piña et al. (2010) argued "*if sulfide segregation had taken place during the emplacement of the Aguablanca stock instead of during the formation of the inferred mafic sequence, Ni–Cu sulfides should be randomly distributed throughout the mafic stock. However, the sulfides are significantly concentrated in those parts of the breccia matrix with the highest density of mafic-ultramafic fragments*". These authors suggested that the presence of Mg-rich olivine with very low Ni contents in peridotite cumulate fragments with further unusually low Cu/Zr ratios (below 1) would be indicative of the crystallization of these peridotite cumulates from a depleted-metal silicate magma due to the early sulfide segregation.

In this scenario, sulfides would have segregated in a staging magma chamber before the fractionation of peridotite cumulates due to addition of crustal sulfur scavenging metals from the silicate magma (Fig. 5.5a). Then, the segregated sulfides would have efficiently separated from the silicate magma and settled toward the base of the chamber (Fig. 5.5b) due to its higher density (~4000 kg/m^3 for sulfide melt, Dobson et al. 2000; ~2600 kg/m^3 for silicate magmas, McBirney and Murase 1984). Above the position where the sulfides accumulated, the metal-depleted silicate magma would have evolved by fractional crystallization giving rise to the sequence of sulfide-free mafic-ultramafic rocks (Fig. 5.5c) that were later disrupted and dispersed as fragments in the breccia (Fig. 5.5d).

However, the mechanism invoked above implies the upward transport of high amounts of dense sulfide melts, and it is unclear which processes can efficiently drive such a process. In their model of emplacement, Tornos et al. (2006) noted this problem and assumed a scenario of overpressured dense magmas channelized upward through small extensional structures within transpressional settings. Similarly, Romeo et al. (2008) suggested multiple magma injections along open tensional cracks related to the Cherneca shear zone. As alternative model, recently, Barnes et al. (2018) have suggested that the sulfide melt may have percolated downward by gravity through a pre-existing, partially molten silicate-matrix intrusion breccia. This model would avoid the problem of upward transport of volumes of sulfide melt with densities greater than the host country rock.

5.3 Fractionation of Sulfide Melt: Origin of the Different Types of Mineralization

Fractional crystallization of sulfide melts is a very common process in sulfide-rich magmatic systems. Experimental studies (Kullerud et al. 1969; Ebel and Naldrett 1996; Barnes et al. 1997, among others) and empirical observations (Zientek et al. 1994; Naldrett et al. 1996) have shown that sulfide melts can differentiate on cooling, crystallizing first a Fe-rich monosulfide solid solution (MSS) at temperatures of 1180–950 °C leaving a Cu-rich residual sulfide melt. The Cu-rich sulfide melt crystallizes at temperatures below 900 °C as intermediate solid solution (ISS, Dutrizac 1976). Below around 650 °C, the MSS decomposes forming pyrrhotite and pentlandite, whereas the ISS recrystallizes as chalcopyrite. The final result is the formation of the sulfide assemblage composed of pyrrhotite, pentlandite and chalcopyrite commonly found in most natural sulfide ores worldwide. In some cases, the Cu-rich

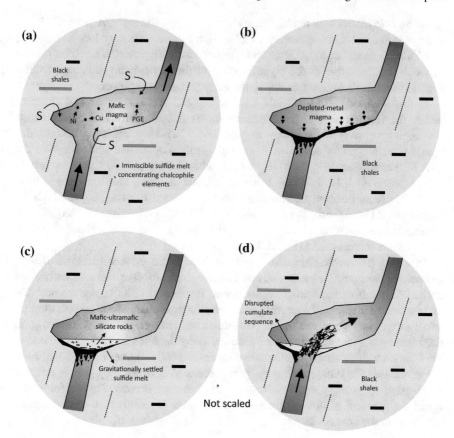

Fig. 5.5 Schematic diagrams illustrating the proposed model by Piña et al. (2010) for the sulfide segregation and later brecciation of the inferred ultramafic sequence in Aguablanca. The shape and size of the drawn chamber is merely schematic. **a** Early sulfide segregation due to the addition of crustal sulfur from the black shales of the Serie Negra Formation scavenging chalcophile elements from the silicate magma; **b** gravitational settling of sulfide melt due to its higher density; **c** the sulfide melt is concentrated on the base of the chamber leaving to an overlying metal-depleted silicate magma which evolved by fractional crystallization giving rise to an inferred magic-ultramafic cumulate sequence; **d** disruption of the mafic-ultramafic sequence and upward transport of silicate fragment along with sulfide and silicate melts probably due to violent melt injections of new fresh silicate magma

sulfide melt can efficiently separate from the MSS, escaping from the solidified ore, and form chalcopyrite-rich veins or stringers adjacent to the MSS (Duran et al. 2017).

During the fractional crystallization of sulfide melts, PGE and other chalcophile elements such as Ni, Co, Se, Bi, As, Te or Au distribute between MSS and Cu-rich sulfide liquid according to their partition coefficients between these phases (Fleet et al. 1993; Li et al. 1996; Barnes et al. 1997; Mungall et al. 2005; Liu and Brenan 2015). Rhenium, Os, Ir, Ru, Rh, and Co partition into the MSS, whereas Pt, Pd, Au,

Ag, Cd and semimetals (As, Bi, Te and As) behave incompatibly and preferentially concentrate in the Cu-rich sulfide liquid. According to this partitioning behaviour, the first elements are preferentially concentrated in those zones of ore enriched into the exsolution products of MSS, i.e. pyrrhotite and pentlandite, whereas that the second elements are preferentially concentrated in the chalcopyrite-rich zones. The partitioning behaviour of Ni between MSS and sulfide melt depends on the temperature and fS_2 of the system (Li et al. 1996; Barnes et al. 1997; Mungall et al. 2005 among others). In general, Ni only shows compatible behaviour into MSS at moderately high sulfur fugacity and low temperatures.

In the Aguablanca ores, the high modal abundance in pyrrhotite of the semi-massive ore (typically >55 vol.% total sulfides), the notable predominance of pentlandite over chalcopyrite (pn/cpy ratios generally >4, and Ni/Cu ratios typically >3), the presence of well developed loop textures with pentlandite around pyrrhotite grain boundaries (formed by high temperature grain boundary exsolution of pentlandite from MSS), and the enrichment in Ni, Os, Ir, Ru and Rh and depletion in Cu, Au, Pt and Pd relative to disseminated ore (Fig. 4.15) suggest that the semi-massive ore represents MSS-enriched cumulates. The compatible behaviour of Ni during MSS formation suggests that MSS fractionation took place under moderately high sulfur fugacity and temperatures not very far from the solidus of the sulfide melt (Mungall et al. 2005). A striking feature of Aguablanca is that although the semi-massive ore seems to represent MSS cumulates and is quite abundant in the deposit, there seems not to be an area in where the Cu-rich sulfide liquid has been efficiently concentrated in form of chalcopyrite-rich veins or stringers. A few chalcopyrite veinlets occur dispersed throughout the deposit. These veinlets clearly represent Cu-rich residual sulfide melts as is indicated by the metal tenors enriched in Cu, Pd, Au and Pt (Fig. 4.13) and the extremely low IPGE and Rh contents of pyrrhotite and pentlandite (Piña et al. 2012). However, they represent less than 1–2 vol.% of the total ore, an amount insignificant considering the high abundance of semi-massive ore present in the deposit. By contrast, chalcopyrite is locally concentrated in Cu-rich semi-massive domains (up to 3.7 wt% Cu) randomly distributed between Ni-rich (Cu-poor) semi-massive sulfides at centimeter scale. Thus, it has been suggested that some amounts of residual Cu-rich sulfide liquid after MSS fractionation may remain retained among crystallizing MSS (Piña et al. 2008) giving rise locally to relatively chalcopyrite-rich semi-massive ore domains. Alternatively, much of the Cu-rich residual sulfide melt may have been moved out of the breccia, possibly by gravity-driven downward percolation. Some evidence for this is seen in the microtextures of the mineralized breccia (Barnes et al. 2018). At mm scale, single-crystal silicate mineral inclusions are preferentially enclosed with pyrrhotite and pentlandite, representing early crystallised cumulus MSS grains, whereas chalcopyrite is relatively poor in silicate inclusions, implying that the Cu-rich sulfide liquid may continue percolating leaving behind a relatively Cu-poor, Ni-enriched MSS component. The implications for exploration are evident. Assuming the model of gravity-driven percolation of sulfide liquid into a pre-existing silicate-matrix breccia, the Cu-rich sulfide liquid could have drained out from the semi-massive sulfides toward at some point below the current breccia, waiting to be discovered at depth.

Fig. 5.6 Metal abundances in pyrrhotite, pentlandite and chalcopyrite from semi-massive ore normalized against the average whole rock content of the disseminated ore (Modified from printed figure by Piña et al. (2012), with permission from Elsevier)

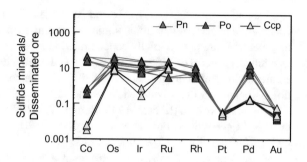

Two different possibilities are considered for the origin of the disseminated ore (Piña et al. 2012): (a) that it represents the in situ crystallization of an original sulfide melt from which MSS formed, or (b) that it represents the crystallization of a fractionated sulfide melt after MSS formation. In both cases, the disseminated sulfides would be impoverished in Ni, Os, Ir, Ru and Rh, and enriched in Cu, Au, Pd, Pt and semimetals in comparison to the semi-massive sulfides such as is observed in the 100% sulfide, mantle-normalized patterns (Fig. 4.15). According to Piña et al. (2012), the latter possibility would be unlikely because the *liquidus* temperature of the silicate magma is higher than that of a fractionated sulfide liquid and thus it would be difficult for a fractionated sulfide melt to disperse to form disseminated sulfides in partially solidified igneous rocks. These authors compared the metal contents in pyrrhotite, pentlandite and chalcopyrite from the semi-massive ore normalized against the average whole rock content of the disseminated ore (Fig. 5.6) and observed that Os, Ir, Ru and Rh are enriched in pyrrhotite and pentlandite by a factor of 3–15 relative to the disseminated ore, whereas Pt and Au are depleted by factors of 0.1 and 0.01. Considering that these values are within the range for the partition coefficients into MSS, the disseminated ore likely represents an original sulfide liquid trapped as droplets among the igneous silicate framework.

The sulfide composition of the Aguablanca disseminated ore can be modelled using the following equation proposed by Campbell and Naldrett (1979):

$$C_i^{sul} = C_i^{sil} * D_i * (R + 1)/(R + D_i) \qquad (5.1)$$

where C_i^{sul} and C_i^{sil} represent the concentrations of the metal i in the sulfide and silicate melt, respectively; D_i is the sulfide/silicate melt partition coefficient of the metal; and R is the R-factor defined as the mass ratio of silicate melt to sulfide melt. It is assumed that the sulfide/silicate melt partition coefficients are 1000 for Cu, 500 for Ni and 500,000 for Pd (Barnes and Ripley 2016, and reference therein). Figure 5.7 illustrating Cu/Pd versus Pd_{100} (100% sulfide) and Pd_{100} versus Ni_{100} shows that the sulfides segregated from a silicate magma containing 250 ppm Cu, 300 ppm Ni and 6 ppb Pd, under R-factors ranging from 200 to 1000, matching quite well with the composition of the Aguablanca disseminated sulfides. These metal values are in the range of values expected for mantle-derived high Mg basaltic magmas. The

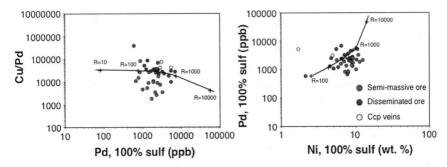

Fig. 5.7 Plots of Cu/Pd versus Pd (100% sulf) (**a**), and Pd (100% sulf) versus Ni (100% sulf) (**b**) for Aguablanca sulfides. Model curves display the compositional variation of sulfide melts segregated from silicate magma containing 250 ppm Cu, 300 ppm Ni and 6 ppb Pd at various R factor values [Data are from Piña et al. (2008) and Peralta (2010)]

estimated value for Cu, 250 ppm, is maybe a bit high considering an arc related mafic magma, but this content may be well reached after 40–50% fractional crystallization of the magma. Additionally, Tornos et al. (2006) suggested that the relatively high Cu content of the Aguablanca sulfides could be due to the assimilation of crustal components hosting small stratabound concentrations of chalcopyrite and bornite.

5.4 Behaviour of PGE

During the fractionation of Aguablanca sulfide melt, Os, Ir, Ru and Rh behaved differently to Pt and Pd according to their different partition coefficients between MSS and sulfide melt ($D^{MSS/sulf}$ ~5–20 for Os, Ir, Ru and Rh, and ~0.1 to $1 \cdot 10^{-3}$ for Pt and Pd, Liu and Brenan 2015). Thus, Os, Ir, Ru and Rh were preferentially concentrated into the MSS, i.e., the semi-massive ore, whereas Pt and Pd remained in the residual sulfide melt. According to the mass balance calculation carried out by Piña et al. (2012), Os, Ir, Ru and, at a lesser extent, Rh are mostly accommodated into pyrrhotite and pentlandite of the semi-massive ore. They are approximately equally distributed between both sulfides, but in detail Os, Ir and Rh are slightly preferentially concentrated in pyrrhotite, reflecting a preference of these elements for pyrrhotite against pentlandite during exsolution processes.

Despite the highest Pd contents occurring in the few chalcopyrite-rich veinlets found in the deposit, in agreement with the incompatible behaviour of Pd during MSS fractionation, it is very probable that some Pd entered MSS, because pentlandite from the semi-massive ore hosts significant amounts of Pd. One might expect that pentlandite exsolved from MSS contains very little Pd, but the true is that the presence of Pd into pentlandite is a common feature in many Ni–Cu–PGE deposits (i.e., Medvezky Creek Mine, Noril'sk, and Merensky Reef, Bushveld Complex, Barnes et al. 2008; J-M Reef, Stillwater Complex, Godel and Barnes 2008; Platreef, Bushveld

Complex, Holwell and McDonald 2007; Creighton deposit, Sudbury, Dare et al. 2010; Fazenda Mirabela intrusion, Knight et al. 2017). Peregoedova and Ohnenstetter (2002) showed that pentlandite can form via exsolution from heazlewoodite-ISS. Since Pd preferentially concentrates into the Cu-rich sulfide melt from which ISS crystallizes, it would be reasonable to find Pd into pentlandite. If this were the case, then the pentlandite should have lower Os, Ir, Ru and Rh than pyrrhotite, because ISS does not concentrate any of these elements. However, in the Aguablanca semi-massive samples, pentlandite has similar amounts of IPGE and Rh to those into pyrrhotite, suggesting that the pentlandite exsolved from MSS. In contrast, the pentlandite present in the chalcopyrite veinlets contains very little IPGE and Rh and has the highest Pd concentrations, suggesting that this pentlandite may have exsolved from ISS on cooling or, at least, from a fractionated sulfide melt depleted in IPGE and Rh by earlier crystallization of MSS.

The origin of Pd in pentlandite has been considered in several studies. Experimental studies at 650 °C in systems with high sulfur fugacity and avoid of semimetals-rich phases (Cafagna and Jugo 2016) have shown that Pd partitions into MSS when the Cu-rich sulfide melt crystallizes in form of ISS. In other words, Pd is not partitioned into the ISS and is concentrated by MSS instead. At high temperatures, MSS can dissolve appreciable amounts of Pd (up to 11 wt% at 900 °C), but with decreasing temperature, the solubility of Pd decreases drastically (0.4 wt% at 500 °C) and as a result Pd enters into the pentlandite structure (Makovicky et al. 1986). In agreement with this experimental evidence, Barnes et al. (2006) and Dare et al. (2010) have attributed the high Pd concentrations in pentlandite from Noril´sk and Sudbury Ni–Cu deposits, respectively, to Pd diffusion into pentlandite from ISS and MSS during cooling. The timing of pentlandite exsolution from MSS and the textural relationships with chalcopyrite (i.e., whether pentlandite is or is not in contact with chalcopyrite) would be key factors in the amount of Pd that enters into pentlandite. Pentlandite starts exsolving from MSS below around 600 °C. At high temperatures, the diffusion rate is high enough for the pentlandite forms large grains at the original MSS grain boundaries, but at lower temperatures (~300 °C) the diffusion rates decrease to the point that Ni can not get far and forms lamellae within pyrrhotite grains (i.e., pentlandite flames) (Kelly and Vaughan 1983). According to Dare et al. (2010)'s diffusion model, the Pd content initially dissolved in MSS/pyrrhotite and ISS would start to be rejected to granular pentlandite at high temperatures (<650 °C). As a consequence, pyrrhotite and ISS would become depleted in Pd and later, at low temperatures (~300 °C), Pd would not be available to diffuse into flame-textured pentlandite. This diffusion model explains the Pd distribution in the Aguablanca pentlandite. The flames of pentlandite are much poorer in Pd than co-existing granular pentlandites, indicating that pyrrhotite had already rejected almost the totality of Pd when the pentlandite flames started to exsolve or the diffusion rate for Pd was too slow at the temperature at which the pentlandite flames formed.

Platinum occurs predominantly in form of PGM, such as is typical in most Ni–Cu–PGE ore deposits (e.g., Dare et al. 2010), and only some minor but significant amounts of Pt have been found in pyrite.

5.5   Origin and Role of Pyrite as Carrier of PGE

Pyrite is a common minor phase in magmatic sulfide ores but, in many cases, its origin is not clear. In general, pyrite is considered a secondary phase formed by replacement of pre-existing sulfides, mostly pyrrhotite and pentlandite, due to the activity of hydrothermal fluids (e.g., Naldrett et al. 1999; Smith et al. 2014; Duran et al. 2015), but it can be also formed by subsolidus reactions involving MSS and ISS on cooling without the involvement of secondary (i.e., hydrothermal) processes (Dare et al. 2011; Cafagna and Jugo 2016).

In Aguablanca, pyrite is particularly abundant, occurring with different textures. Its origin and relationship with the other sulfides has become important from Piña et al. (2012)'s study who showed that pyrite hosts significant amounts of PGE, specially Rh and Pt. Pyrite replacing plagioclase is considered the result of the activity of late magmatic/hydrothermal fluids (Piña et al. 2013), but the origin of the pyrites texturally related to sulfides is much less clear. According to Piña et al. (2013), all pyrites independently their textural variety would be genetically related to the circulation of magmatic/hydrothermal fluids. Among other evidence, they based their interpretation in the fact that with the exception of PGE and other metals (Ni, Co, Ag), all pyrites (including that replacing plagioclase) share similar concentrations in As, Bi, Te, Sb, Se and some mobile metals such as Au and Cu (Fig. 4.18). If pyrite texturally related to sulfides were primary in origin (i.e., subsolidus product from MSS and/or ISS), then the trace element contents should be quite different to those of the pyrite replacing plagioclase and this is not the case. An increase in the fS_2 and fO_2 of the late magmatic/hydrothermal fluids were probably the factors triggering the destabilization of pyrrhotite in favour of pyrite.

In Aguablanca, pyrite contains similar amounts of Os, Ir, Ru and Rh to those in pyrrhotite; negligible Pd contents and, interestingly, is the unique sulfide containing appreciable amounts of Pt. The PGE distribution in the different types of pyrite is result of the combination of two factors (Piña et al. 2013): (a) the abundance of PGE in the original minerals, and (b) the mobile/immobile behaviour of PGE during the circulation of fluids responsible of replacement of pyrrhotite and plagioclase by pyrite. During the replacement of pyrrhotite by pyrite, highly immobile elements such as Os, Ir, Ru and Rh were rapidly incorporated into pyrite, which thus inherited the IPGE and Rh contents of the replaced pyrrhotite. By contrast, other mobile elements such as Au, Bi, Te, Sb and As could not be inherited by pyrite from pyrrhotite because pyrrhotite does not host these elements and probably were incorporated into pyrite via fluids (Piña et al. 2013).

Platinum is not present in pyrrhotite, and consequently most pyrite does not contain Pt. However, some large idiomorphic pyrite grains contain relatively high Pt contents as illustrated in the compositional map of Fig. 4.19b. Platinum is a mobile element that can be introduced in the system by hydrothermal fluids. Alternatively, Piña et al. (2013) have suggested that the local enrichment in Pt of some large idiomorphic pyrite grains may be due to the dissolution of discrete PGM, particularly sperrylite, by the same fluids responsible of pyrrhotite replacement by pyrite, making Pt available for incorporation into pyrite. This mechanism of release of Pt from sperrylite during late alteration has been also invoked by Suárez et al. (2010) to explain the presence of Pt into secondary minerals (e.g., goethite) in the Aguablanca gossan.

5.6   Origin of Platinum-Group Minerals

The PGM assemblage at Aguablanca is dominated by (Pd–Pt–Ni) bismuthotellurides (merenskyite, palladian melonite, michenerite and moncheite) and Pt arsenides (sperrylite) texturally related to the base metal sulfides. Two different possibilities are commonly considered for the origin of Pd–Pt bismuthotellurides within magmatic sulfides: they can form by exsolution from base metal sulfides during cooling or can crystallize directly from a highly fractionated PGE- and semimetal-rich melt (Helmy et al. 2007; Tomkins 2010; Holwell and McDonald 2010). Experimental studies have shown that Pd, Pt and semimetals (Bi, Te, Sb, As), rather than partitioning into ISS, are preferentially concentrated in an immiscible semimetal rich melt, which remains liquid when ISS has already crystallized (Liu and Brenan 2015). Pd–Pt bismuthotellurides can then crystallize from this melt preferentially along grain boundaries of sulfide and silicate. At Aguablanca, the origin of PGM has been in detail discussed by Ortega et al. (2004) and Piña et al. (2008, 2012). In these studies, it has been suggested that the (Pd–Pt–Ni) bismuthotellurides were formed during the low-temperature re-equilibration of the ore. They based an origin of bismuthotellurides as exsolutions from the base metal sulfides in the following lines of evidence: (a) most of the PGM are included within individual sulfides, rather than around the margins of sulfides which would be more common if PGM had crystallized from an immiscible semimetal-rich melt (e.g., Tomkins 2010); (b) the (Pd–Pt–Ni) bismuthotellurides show textures also observed in PGM formed experimentally by exsolution from MSS (Peregoedova et al. 2004); and (c) the presence of tiny Pd–Pt–Bi–Te microinclusions revealed during laser ablation analyses that points to exsolution processes during cooling. Furthermore, it is unlikely that the amount of Te in the Aguablanca sulfide melt was abundant enough to segregate a Te-rich melt. Helmy et al. (2007) experimentally showed that it is necessary that the concentration of Te in the sulfide melt exceeds its solubility in MSS and ISS, ~0.2 wt%, to segregate a Te-rich melt. At Aguablanca, if all the Te present in the mineralized samples were assigned to the sulfides, then the sulfide liquid would only have ~15 ppm. This amount of Te can easily dissolve in the sulfides and thus is difficult to infer the immiscibility of a Te-rich melt. In addition, the relatively high Pd contents in pentlandite from semi-massive sulfides are in agreement with the absence of an immiscible semimetal rich melt, since then Pd would have been preferentially concentrated into the semimetal rich liquid and strongly depleted into the MSS (i.e., Cafagna and Jugo 2016). Thus, it seems probable that Pd and Pt as well as Te, Bi and As, initially in pyrrhotite and pentlandite, were no longer able to remain dissolved in the sulfides upon cooling and were expelled giving rise to the Aguablanca (Pd–Pt–Ni) bismuthotellurides assemblage. This scenario also implies that much of Pt, Pd, Te, Bi and As, remained in the MSS during the early stages of cooling, probably, as a consequence of the rapid cooling of the ore that also limited the extensive fractionation of the sulfide melt.

5.7 Emplacement of the Aguablanca Ore Deposit

The structural history of the Aguablanca stock and the cross section geometry based on the 3D gravity modelling led Romeo et al. (2008) to conclude that the Aguablanca stock has an inverted drop geometry with the root of the intrusion located in its northern margin adjacent to the Cherneca ductile shear zone (Fig. 3.10). The long axis of this root in the 3D gravity model is not parallel but is also oblique (~50°) with respect to the Cherneca shear zone (Fig. 3.11). This oblique orientation fits well with that expected for tensional fractures developed within a transpressional sinistral strike-slip shear structure corresponding to the Cherneca shear zone (Fig. 3.11). Thus, Romeo et al. (2008) have proposed that the Aguablanca stock was emplaced due to successive opening events of these tensional fractures related to the Cherneca shear zone.

Tornos et al. (2006) and Piña et al. (2010) envisaged the emplacement of the Aguablanca orebodies through multiple injections of overpressured magma mainly controlled by the gradual opening of tensional fractures. According to this model, overpressure would have been the responsible of brecciation and transport upward of the dense sulfide- and fragment-charged silicate magmas from a deep level to a shallower site through low-density crustal rocks. The unmineralized igneous rocks of the Aguablanca stock (mostly, gabbrodiorites) would represent early injections of fractionated silicate melt from the uppermost parts of the deep magma chamber, whereas semi-massive ore would represent the latest injections containing a mixture of sulfide melt, partially consolidate mafic-ultramafic fragments and remaining silicate melt. The emplacement model envisaged by Tornos et al (2001) and Piña et al. (2010) is described as following. The first stage is the injection of sulfide-free differentiated silicate melts carrying minor mafic-ultramafic fragments. These melts would have flowed toward the south, taking on the inverted drop geometry of the Aguablanca stock and would have evolved by fractional crystallization giving rise to sulfide-free gabbronorite, norite and gabbrodiorite. Later, a new injection of silicate magma coming from the same magmatic source but containing droplets of unfractionated sulfide melt would have generated the disseminated ore-bearing gabbronorites and norites, likely accreting inward from the sidewalls of the Aguablanca rocks. Tornos et al. (2006) suggested that the barren igneous rocks and the disseminated ore-bearing rocks may represent the product of turbulent and variable mixing between fragments and sulfide and silicate melts. Finally, the emplacement of the semi-massive ore would have taken place in the latest stages of emplacement of the intrusion. In this last injection, the sulfide-rich silicate melt charged with mafic-ultramafic fragments from deep, partially consolidated rocks, would have given rise to the semi-massive ore in the inner parts of the mineralized breccia.

An alternative model has been recently proposed by Barnes et al. (2018) based on the textures observed in the Aguablanca breccias, where sulfide melt is seen to flood the pore space between melting and disaggregating silicate rock fragments by a process analogous to that responsible for similar sulfide matrix ore breccias at Voisey's Bay (Barnes et al. 2017b). In this model, the disseminated ore represents an early stage of emplacement developed as sulfide-bearing cumulates within a funnel-shaped widening of the Aguablanca subvertical feeder. According to these authors, the semi-massive ore-bearing breccia can have formed by a number of mechanisms. A possibility is that the breccia represents a downward-directed gravity flow of sulfide melt, silicate phenocrysts and igneous and country rock fragments developed during a late drain-back stage as the magma flux through the intrusion network waned. Alternatively, an upward-migrating "sludge" of silicate melt, crystals and rock fragments may also have collapsed due to a great increase of the transported load to the point where the suspension was no longer buoyant relative to the country rocks, forming a "log jam" that choked the flow into the magma chamber. The sulfide melt may have deposited initially in gravitationally unstable pools higher in intrusion network of which the Aguablanca stock is a part, and have then percolated downward through the previously emplaced, partially molten, silicate-matrix breccia. Superheat introduced into the breccia by sulfide liquid percolation led to melting and displacement of the original interstitial silicate melt, leaving behind refractory rock fragments, pyroxene crystals and an interstitial network of sulfide melt.

References

Barnes S-J, Ripley EM (2016) Highly siderophile and strongly chalcophile elements in magmatic ore deposits. Rev Mineral Geochem 81:725–774

Barnes S-J, Boyd R, Kornelliussen A, Nilssen LP, Often M, Pedersen RB, Robins B (1987) The use of noble and base metal ratios in the interpretation of ultramafic and mafic rocks, examples from Norway. Geoplatinum 87 Symposium Open University, Milton Keynes, April

Barnes S-J, Makovicky E, Karup-Moller S, Makovicky M, Rose-Hanson J (1997) Partition coefficients for Ni, Cu, Pd, Pt, Rh and Ir between monosulfide solid solution and sulfide liquid and the implications for the formation of compositionally zoned Ni–Cu sulfide bodies by fractional crystallization of sulfide liquid. Can J Earth Sci 34:366–374

Barnes S-J, Melezhik VA, Sokolov SV (2001) The composition and mode of formation of the Pechenga nickel deposits, Kola Peninsula, northwestern Russia. Can Mineral 39:447–471

Barnes S-J, Cox RA, Zientek ML (2006) Platinum-group element, gold, silver and base metal distribution in compositionally zoned sulfide droplets from the Medvezky Creek Mine, Norilsk, Russia. Contrib Mineral Petrol 152:187–200

Barnes S-J, Prichard HM, Cox RA, Fisher PC, Godel B (2008) The location of the chalcophile and siderophile elements in platinum group element ore deposits (a textural, microbeam and whole rock geochemical study): implications for the formation of the deposits. Chem Geol 248:295–317

Barnes SJ, Holwell DA, LeVaillant M (2017a) Magmatic sulfide ore deposits. Elements 13:89–95

Barnes SJ, Le Vaillant M, Lightfoot PC (2017b) Textural development in sulfide-matrix ore breccias in the Voisey's Bay Ni–Cu–Co deposit, Labrador, Canada. Ore Geol Rev 90:414–438

Barnes SJ, Piña R, Le Vaillant M (2018) Textural development in sulfide-matrix ore breccias in the Aguablanca Ni–Cu deposit, Spain, revealed by X-ray fluorescence microscopy. Ore Geol Rev 95: 849–862

Cafagna F, Jugo PJ (2016) An experimental study on the geochemical behaviour of highly siderophile elements (HSE) and metalloids (As, Se, Sb, Te, Bi) in a mss-iss-pyrite system at 650 °C: A possible magmatic origin for Co-HSE-bearing pyrite and the role of metalloid-rich phases in the fractionation of HSE. Geochim Cosmochim Acta 178: 233–258

Campbell IH, Naldrett AJ (1979) The influence of silicate: sulphide ratios on the geochemistry of magmatic sulphides. Econ Geol 74:503–1505

Casquet C, Eguiluz L, Galindo C, Tornos F, Velasco F (1998) The Aguablanca Cu–Ni–(PGE) intraplutonic ore deposit (Extremadura, Spain). Isotope (Sr, Nd, S) constraints on the source and evolution of magmas and sulfides. Geogaceta 24:71–74

Cowden A, Donaldson MJ, Naldrett AJ, Campbell IH (1986) Platinum-group elements in the komatiite-hosted Fe–Ni–Cu sulfide deposits at Kambalda Western Australia. Econ Geol 81:1226–1235

Dare SAS, Barnes S-J, Prichard H (2010) The distribution of platinum group elements (PGE) and other chalcophile elements among sulfides from the Creighton Ni–Cu–PGE sulfide deposit, Sudbury, Canada, and the origin of palladium in pentlandite. Miner Deposita 45:765–793

Dare SAS, Barnes S-J, Prichard H, Fisher PC (2011) Chalcophile and platinum-group element (PGE) concentrations in the sulfide minerals from the McCreedy East deposit, Sudbury, Canada, and the origin of PGE in pyrite. Miner Deposita 46:381–407

Dobson DP, Crichton WA, Vocadlo L, Jones AP, Wang Y, Uchida T, Rivers M, Sutton S, Brodholt JP (2000) In situ measurement of viscosity of liquids in the Fe–FeS system at high pressures and temperatures. Am Mineral 85:1838–1842

Duran CJ, Barnes S-J, Corkery JT (2015) Chalcophile and platinum group element distribution in pyrites from the sulfide-rich pods of the Lac des Iles Pd deposits, Western Ontario, Canada: implications for post-cumulus re-equilibration of the ore and the use of pyrite composition exploration. J Geochem Explor 158:223–242

Duran CJ, Barnes S-J, Plese P, Kudrna Prasek M, Zientek ML, Pagé P (2017) Fractional crystallization-induced variations in sulfides from the Noril'sk-Talnakh mining district (polar Siberia, Russia). Ore Geol Rev 90:326–351

Dutrizac JE (1976) Reactions in cubanite and chalcopyrite. Can Mineral 14:172–181

Ebel D, Naldrett AJ (1996) Experimental fractional crystallization of Cu- and Ni-bearing Fe-sulfide liquids. Econ Geol 91:607–621

Eckstrand OR, Hulbert LJ (1987) Selenium and the source of sulfur in magmatic nickel and platinum deposits. GAC-MAC Jt Annu Meet Program Abs 12:4

Fleet ME, Chryssoulis SL, Stone WE, Weisener CG (1993) Partitioning of platinum-group elements and Au in the Fe–Ni–Cu–S system: experiments on the fractional crystallization of sulfide melt. Contrib Mineral Petrol 115:36–44

Godel B, Barnes SJ (2008) Platinum-group elements in sulfide minerals and the whole rocks of the J-M Reef (Stillwater Complex): implication for the formation of the reef. Chem Geol 248:272–294

Helmy HM, Ballhaus C, Berndt J, Bockrath C, Wohlgemuth-Ueewasser C (2007) Formation of Pt, Pd and Ni tellurides: experiments in sulphide-telluride systems. Contrib Mineral Petrol 153:577–591

Holwell DA, McDonald I (2007) Distribution of platinum-group elements in the Platreef at Overysel, northern Bushveld Complex: a combined PGM and LA-ICP-MS study. Contrib Mineral Petrol 154:171–190

Holwell DA, McDonald I (2010) A review of the behaviour of platinum group elements within natural magmatic sulfide ore systems: the importance of semimetals in governing partitioning behaviour. Platin Met Rev 54:26–36

Howard JH (1977) Geochemistry of selenium: formation of ferroselite and selenium behavior in the vicinity of oxidizing sulfide and uranium deposits. Geochim Cosmochim Acta 41:1665–1678

Keays RR (1995) The role of komatiitic and picritic magmatism and S-saturation in the formation of the ore deposits. Lithos 34:1–18

Keays RR, Lightfoot PC (2010) Crustal sulfur is required to form magmatic Ni–Cu sulfide deposits: Evidence from chalcophile element signatures of Siberian and Deccan Trap basalts. Miner Deposita 45:241–257

Kelly DP, Vaughan DJ (1983) Pyrrhotite–pentlandite ore textures: a mechanistic approach. Min Mag 47:453–463

Knight RD, Prichard HM, Filho CF (2017) Evidence for as contamination and the partitioning of Pd into pentlandite and platinum group elements into pyrite in the Fazenda Mirabela Intrusion, Brazil. Econ Geol 112:1889–1912

Kullerud G, Yund RA, Moh G (1969) Phase relations in the Fe–Ni–S, Cu–Fe–S and Cu–Ni–S systems. Econ Geol Monogr 4:323–343

Lesher CM, Burnham OM (2001) Multicomponent elemental and isotopic mixing in Ni–Cu–(PGE) ores at Kambalda, western Australia. Can Mineral 39:421–446

Li C, Naldrett AJ (1993) Sulfide capacity of magma: a quantitative model in its application to the formation of sulfide ores at Sudbury, Ontario. Econ Geol 88:1253–1260

Li C, Barnes S-J, Makovicky E, Rose-Hansen J, Makovicky M (1996) Partitioning of Ni, Cu, Ir, Rh, Pt and Pd between monosulfide solid solution and sulfide liquid: effects of composition and temperature. Geochim Cosmochim Acta 60:1231–1238

Liu Y, Brenan J (2015) Partitioning of platinum-group elements (PGE) and chalcogens (Se, Te, As, Sb, Bi) between monosulfide-solid solution (MSS), intermediate solid solution (ISS) and sulfide liquid at controlled fO2–fS2 conditions. Geochim Cosmochim Acta 159:139–161

Maier WD, Barnes S-J (2010) The Kabanga Ni sulfide deposits, Tanzania: II. Chalcophile and sidérophile element geochemistry. Miner Deposita 45:443–460

Makovicky M, Makovicky E, Rose-Hansen J (1986) Experimental studies on the solubility and distribution of platinum group elements in base-metal sulfides in platinum deposits. In Gallagher MJ, Ixer RA, Neary CR, Prichard HM (eds) Metallogeny of basic and ultrabasic rocks, Institute of Mining and Metallurgical Special Publication, pp 415–426

Mathur R, Tornos F, Barra F (2008) The Aguablanca Ni–Cu deposit: a Re-Os isotope study. Int Geol Rev 50:948–958

McBirney AR, Murase T (1984) Rheological properties of magmas. Annu Rev Earth Planet Sci 12:337–357

McDonough WF, Sun SS (1995) The composition of the Earth. Chem Geol 120:223–253

McGoldrick PJ, Keays RR (1981) Precious and volatile metals in the perseverance nickel deposit gossan: implications for exploration in weathered terrains. Econ Geol 76:1752–1763

Mungall JE, Andrews R, Cabri LJ, Sylvester PJ, Tubrett M (2005) Partitioning of Cu, Ni, An, and platinum-group elements between monosulfide solid solution and sulfide melt under controlled oxygen and sulfur fugacities. Geochim Cosmochim Acta 69:4349–4360

Naldrett AJ (2004) Magmatic sulfide deposit: geology, geochemistry and exploration. Springer, Berlin, p 728

Naldrett AJ, Craig JR, Kullerud G (1967) The central portion of the Fe–Ni–S system and its bearing on pentlandite solution in iron-nickel sulfide ores. Econ Geol 62:826–847

Naldrett AJ, Innes DG, Sowa J, Gorton M (1982) Compositional variation within and between five Sudbury ore deposits. Econ Geol 77:1519–1534

Naldrett AJ, Fedorenko AV, Asif M, Lin S, Kunilov VE, Stekhin AI, Lightfoot PC, Gorbachev NS (1996) Controls on the composition of Ni–Cu sulfide deposit as illustrated by those at Noril'sk, Siberia. Econ Geol 91:751–773

Naldrett AJ, Asif M, Schandl E, Searcy T, Morrison GG, Binney WP, Moore C (1999) Platinum-group elements in the Sudbury ores: significance with respect to the origin of different ore zones and to the exploration for footwall orebodies. Econ Geol 94:185–210

Ohmoto H (1986) Stable isotope geochemistry of ore deposits. Rev Mineral Geochem 16:491–559

Ortega L, Lunar R, García Palomero F, Moreno T, Martín Estévez JR, Prichard HM, Fisher PC (2004) The Aguablanca Ni–Cu–PGE deposit, southwestern Iberia: magmatic ore-forming processes and retrograde evolution. Can Mineral 42:325–350

Peach CL, Mathez EA, Keays RR (1990) Sulfide melt-silicate melt distribution coefficients for noble metals and other chalcophile elements as deduced from MORB: implications for partial melting. Geochim Cosmochim Acta 54:3379–3389

Peregoedova AV, Ohnenstetter M (2002) Collectors of Pt, Pd and Rh in a S-poor Fe–Ni–Cu-sulfide system at 760 °C: results of experiments and implications for natural systems. Chem Geol 208:247–264

Peregoedova AV, Barnes S-J, Baker DR (2004) The formation of Pt–Ir alloys and Cu–Pd-rich sulfide melts by partial desulfurization of Fe–Ni–Cu sulfides: Results of experiments and implications for natural systems. Chem Geol 208:247–264

Pereira MF, Chichorro M, Linnemann U, Eguiluz L, Silva B (2006) Inherited arc signature in Ediacaran and Early Cambrian basins of the Ossa-Morena Zone (Iberian Massif, Portugal): paleogeographic link with European and North African Cadomian correlatives. Precambrian Res 144:297–315

Piña R (2006) El yacimiento de Ni–Cu–EGP de Aguablanca (Badajoz): Caracterización y modelización metalogenética. PhD thesis, Universidad Complutense de Madrid, Spain, p 254

Piña R, Lunar R, Ortega L, Gervilla F, Alapieti T, Martínez C (2006) Petrology and geochemistry of mafic-ultramafic fragments from the Aguablanca (SW Spain) Ni–Cu ore breccia: Implications for the genesis of the deposit. Econ Geol 101:865–881

Piña R, Gervilla F, Ortega L, Lunar R (2008) Mineralogy and geochemistry of platinum-group elements in the Aguablanca Ni–Cu deposit (SW Spain). Miner Petrol 92:259–282

Piña R, Romeo I, Ortega L, Lunar R, Capote R, Gervilla F, Tejero R, Quesada C (2010) Origin and emplacement of the Aguablanca magmatic Ni–Cu–(PGE) sulfide deposit, SW Iberia: a multidisciplinary approach. Geol Soc Am Bull 122:915–925

Piña R, Gervilla F, Barnes S-J, Ortega L, Lunar R (2012) Distribution of platinum-group and chalcophile elements in the Aguablanca Ni–Cu sulfide deposit (SW Spain): evidence from a LA-ICP-MS study. Chem Geol 302–303:61–75

Piña R, Gervilla F, Barnes S-J, Ortega L, Lunar R (2013) Platinum-group elements-bearing pyrite from the Aguablanca Ni–Cu sulphide deposit (SW Spain): a LA-ICP-MS study. Eur J Mineral 25:241–252

Queffurus M, Barnes S-J (2015) A review of sulfur to selenium ratios in magmatic nickel–copper and platinum-group element deposits. Ore Geol Rev 69:301–324

Ripley EM (1990) Se/S ratios of the Virginia formation and Cu–Ni mineralization in the Babbitt area, Duluth Complex, Minnesota. Econ Geol 85:1935–1940

Ripley EM, Li C (2013) Sulfide saturation in Mafic Magmas: is external sulfur required for magmatic Ni–Cu–(PGE) ore genesis? Econ Geol 108:45–58

Robertson J, Ripley EM, Barnes SJ, Li C (2015) Sulfur liberation from country rocks and incorporation in mafic magmas. Econ Geol 110:1111–1123

Romeo I, Tejero R, Capote R, Lunar R (2008) 3-D gravity modelling of the Aguablanca Stock, tectonic control and emplacement of a Variscan gabbronorite bearing a Ni–Cu–PGE ore, SW Iberia. Geol Mag 145:345–359

Nägler T (1990) Sm-Nd, Rb-Sr and common lead isotope geochemistry on fine-grained sediments of the Iberian Massif. PhD thesis, ETH, Zurich, p 141

Peralta A (2010) Estudio mineralógico y geoquímico del cuerpo profundo del yacimiento de Ni–Cu-EGP de Aguablanca. MSc. Universidad de Granada, Spain, p 52

Samalens N, Barnes S-J, Sawyer EW (2017) The role of black shales as a source of sulfur and semimetals in magmatic nickel-copper deposits: example from the Partridge River Intrusion, Duluth Complex, Minnesota, USA. Ore Geol Rev 81:173–187

Smith JW, Holwell DA, McDonald I (2014) Precious and base metal geochemistry and mineralogy of the Grasvally Norite-Pyroxenite-Anorthosite (GNPA) member, northern Bushveld Complex, South Africa: implications for a multistage emplacement. Miner Deposita 49:667–692

Song XY, Li XR (2009) Geochemistry of the Kalatongke Ni–Cu–(PGE) sulfide deposit, NW China: implications for the formation of magmatic sulfide mineralization in a postcollisional environment. Miner Deposita 44:303–327

Su S, Li C, Zhou MF, Ripley EM, Qi L (2008) Controls on variations of platinum-group element concentrations in the sulfide ores of the Jinchuan Ni–Cu deposit, western China. Miner Deposita 43:609–622

Suárez S, Prichard HM, Velasco F, Fisher PC, McDonald I (2010) Alteration of platinum-group minerals and dispersion of platinum-group elements during progressive weathering of the Aguablanca Ni–Cu deposit, SW Spain. Miner Deposita 45:331–350

Sun SS, McDonough WF (1989) Chemical and isotopic systematics of oceanic basalts: implication for mantle composition and processes. In: Saunders AD, Norry MJ (eds) Magmatism in the ocean basins, Geological Society London Special Publication, p 313–345

Thériault RD, Barnes S-J (1998) Compositional variations in Cu-Ni-PGE sulfides of the Dunka Road deposit, Duluth Complex, Minnesota: the importance of combined assimilation and magmatic processes. Can Mineral 36:869–886

Tomkins AG (2010) Wetting facilitates late-stage segregation of precious metal–enriched sulfosalt melt in magmatic sulfide systems. Geology 38:951–954

Tornos F, Velasco F (2002) The Sultana orebody (Ossa Morena Zone, Spain): insights into the evolution of Cu–(Au–Bi) mesothermal mineralization. In: Blundell DJ (ed) GEODE Study Centre, Grenoble, p 17

Tornos F, Chiaradia M (2004) Plumbotectonic evolution of the Ossa-Morena Zone, Iberian Peninsula: tracing the influence of mantle-crust interaction in ore-forming processes. Econ Geol 99:965–985

Tornos F, Casquet C, Galindo C, Velasco F, Canales A (2001) A new style of Ni–Cu mineralization related to magmatic breccia pipes in a transpressional magmatic arc, Aguablanca, Spain. Miner Deposita 36:700–706

Tornos F, Inverno C, Casquet C, Mateus A, Ortiz G, Oliveira V (2004) The metallogenic evolution of the Ossa Morena Zone. J Iber Geol 30:143–180

Tornos F, Galindo C, Casquet C, Rodríguez Pevida L, Martínez C, Martínez E, Velasco F, Iriondo A (2006) The Aguablanca Ni-(Cu) sulfide deposit, SW Spain: geologic and geochemical controls and the relationship with a midcrustal layered mafic complex. Miner Deposita 41:737–769

Zientek ML, Likhachev AP, Kunilov VE, Barnes SJ, Meier AL, Carlson RR, Briggs PH, Fries TL, Adrian BM (1994) Cumulus processes and the composition of magmatic ore deposits: examples from the Talnakh district, Russia. Ontario Geol Surv Spec Publ 5:373–392